Applications of Measure Theory to Statistics

Gogi Pantsulaia

Applications of Measure Theory to Statistics

Springer

Gogi Pantsulaia
Department of Mathematics
Georgian Technical University
Tbilisi
Georgia

ISBN 978-3-319-83322-4 ISBN 978-3-319-45578-5 (eBook)
DOI 10.1007/978-3-319-45578-5

Softcover reprint of the hardcover 1st edition 2017

Printed on acid-free paper

This Springer imprint is published by Springer Nature
The registered company is Springer International Publishing AG
The registered company address is: Gewerbestrasse 11, 6330 Cham, Switzerland

Contents

About the Author

Prof. Gogi Pantsulaia graduated in Mathematics from the Iv. Javakhishvili Tbilisi State University (Georgia) in 1982, and received his Ph.D. in mathematics at the Institute of Mathematics of the Ukrainian Academy of Sciences (Ukraine) in 1985. In 2003 he received his degree of doctor of physics and mathematics from the I. Vekua Institute of Applied Mathematics of the Iv. Javakhishvili Tbilisi State University (Georgia). He is a full professor in the Department of Mathematics of the Georgian Technical University (Georgia). He has participated in more than 10 research projects and is the author of four books and more than 75 papers. He has also supervised several doctoral theses. His current research interests include set theory, measure theory, probability theory, and mathematical statistics.

About the Author

Introduction

As Robinson and Wainer [RW] have observed in the almost 300 years since its introduction by Arbuthnot [A], null hypothesis significance testing (NHST) has become an important tool for working scientists. In the early 20th century, the founders of modern statistics (R.A. Fisher, Jerzy Neyman, and Egon Pearson) showed how to apply this tool in widely varying circumstances, often in agriculture, that were almost all very far afield from Dr. Arbuthnot's noble attempt to prove the existence of God.

In the process of applying NHST, however, confusion has often arisen among practitioners and statisticians, giving rise to philosophical criticisms of hypothesis texting (see, for example, [Ch, LMS, K, MZ, MH, O], which cite 300–400 primary references). These criticisms reflected a general point of view that the theory of mathematical statistics and the results of testing are inconsistent in many situations and that the typical null hypothesis is almost always false. With the advantage of increasing use, practitioner's eyes became accustomed to the darker reality and the shortcomings of NHST became more apparent.

Below we cite comments concerning the criticism of NHST from the joint work of J.L. Rodgers and D.C. Rowe [RR]:

> The methodological literature contains cogent criticism of null hypothesis significance testing (NHST; e.g., Cohen, 1994; Rozeboom, 1960; Schmidt, 1996), including the extreme position that NHST has been discredited and never makes a positive contribution (Schmidt Hunter, 1997, p. 37). Some have called for NHST to be outlawed, with enforcement charged to journal editors (see discussion in Shrout, 1997). The American Psychological Association (APA) Board of Scientific Affairs commissioned a task force on statistical inference, a committee of talented methodological scholars, to consider the situation. Their conclusion (Wilkinson Task Force on Statistical Inference, APA Science Directorate, 1999) was to temper the stridency of the anti-NHST movement:
>
> Some had hoped that this task force would vote to recommend an outright ban on the use of significance tests in psychology journals. . . . the task force hopes instead that this report will induce editors, reviewers, and authors to recognize practices that institutionalize the thoughtless application of statistical methods. (pp. 602–603).

> Most thoughtful methodologists seem to conclude that NHST is a statistical tool that handles a certain type of question (e.g., Abelson, 1997; Muliak, Raju, Harshman, 1997; Wainer, 1999). As with any tool, problems can arise with misuse, and other tools can substitute for or complement NHST. However, the committee and many others concluded that NHST should remain part of the methodologists toolbox, along with confidence intervals, effect sizes, graphical analysis, and so on. We were reminded of this recent NHST controversy when we read Roberts and Pashler's (2000) article, which criticized the evaluation of how well a model fits empirical data in the development of psychological and social science theories. Like the critics of NHST, Roberts and Pashler had no difficulty identifying examples in which model-fitting methodology has been misused. Like the critics of NHST, they also pointed to methods to evaluate models that can substitute or complement the evaluation of goodness-of-fit procedures. Like the critics of NHST, they substantially overstated their case in their criticism of using good fits to support theories (p. 365).

On the one hand, the reexamination of the viability of NHST was described by Anderson, Burnham, and Thompson (2000), who showed that over the past 60 years an increasing number of articles have questioned the utility of NHST.

On the other hand, it is revealing that Thompson's database, over the same time period, showed a concomitant increase in the number of articles defending the utility of NHST.

It is obvious that this phenomenon can formally be explained as follows:

- An application of NHST has a supporter (or an oppositionist) if the associated statistical test is "objective" (or "subjective").

One of the purposes of this book consists in putting into notions "objective" and "subjective" reasonable mathematical senses and in providing this simple explanation with a strong mathematical base.

It seems worthwhile also to use "subjective" statistical tests to try to construct new "objective" statistical tests under which NHST remains a viable tool. Here we will present a methodology for resolution of this problem under some restrictions introduced by Pantsulaia and Kintsurashvili [PK2].

We are not going to consider in detail all relevant issues. Instead we shall focus our attention on a certain confusion which is described in the works of Nunnally [N] and Cohen [Coh].

In 1960, Nunnally [N] noticed that in many standard statistical tests null hypotheses are always rejected and observed: "If the decisions are based on convention they are termed arbitrary or mindless while those not so based may be termed subjective. To minimize type *II* errors, large samples are recommended. In psychology practically all null hypotheses are claimed to be false for sufficiently large samples so … it is usually nonsensical to perform an experiment with the sole aim of rejecting the null hypothesis".

In 1994, Cohen [Coh] noticed some gaps between the theory of mathematical statistics and the results of testing and observed: "… Don't look for a magic alternative to NHST [null hypothesis significance testing] … It does not exist."

Here the following question naturally arises:

- *Whether can be explained Jacob Cohen and Jum Nunnally above mentioned observations?*

Estimating a useful signal for a linear one-dimensional stochastic system, we plan to demonstrate the validity of Cohen and Nunnally's predictions for a certain standard hypothesis testing in terms of infinite samples such that the sum of errors of *I* and *II* types is equal to zero (we refer to such tests as tests with a maximal reliability). Note that working with infinite samples is a natural requirement because a definition of consistent estimates can not be given without infinite samples. Further, we plan to explain why a null hypothesis is claimed to be false for "almost every" (in the sense of [HSY]) infinite sample.

Another goal of the present book is an application of the approach of "almost every" (in the sense of [HSY]) in studying structures of domains of some infinite sample statistics and in explaining why the null hypothesis is rejected for "almost every" (in the sense of [HSY]) infinite sample by the associated NHST with a maximal reliability.

In order to explain the large gap between the theory of mathematical statistics and the results of hypothesis testing, by using the technique of Haar null sets in the space of infinite samples, we introduce an essentially new approach which naturally divides the class of all consistent infinite sample estimates of a useful signal in the linear one-dimensional stochastic model into disjoint classes of subjective and objective estimates. Following this approach, each consistent infinite sample estimate has to pass a theoretical test on objectivity. This means that theoretical statisticians should expend much effort in carrying out such a certification exam for each consistent infinite-sample estimation.

Correspondingly, we have the following three objectives:

i. To introduce a new approach which naturally divides the class of all consistent estimates of an unknown parameter in a Polish group into disjoint classes of subjective and objective estimates.
ii. To construct tests on objectivity for consistent estimations of an unknown parameter in a Polish group.
iii. To explain of the main requirement why each consistent infinite-sample estimation must pass the certification exam on objectivity.

This book is devoted to the mathematical development of the first two items. We also briefly discuss a few interesting mathematical points concerning item (iii) (While there is a rich family of NHTSs whose corresponding statistical tests are consistent, to date we have no information regarding their objectivity).

The book comprises six chapters, as outlined below:

Chapter 1 demonstrates that the technique for numerical calculation of some one-dimensional improper Riemann integrals is similar to the technique given by Weyl's [W] celebrated theorem for continuous functions on $[0, 1]$.

In Sect. 1.2 we consider some auxiliary notions and facts from the theory of uniformly distributed sequences on the interval $[0, 1]$. In Sect. 1.3 we present the proof of a certain modification of Kolmogorov's Strong Law of Large Numbers and the Glivenko-Cantelli theorem. In Sect. 1.4 we give an extension of the main result of Baxa and Schoiβengeier [BS] for calculation of some improper one-dimensional Riemann integrals by use of uniformly distributed sequences.

Chapter 2 presents a concept of infinite-dimensional Monte-Carlo integration developed by Pantsulaia [P6].

In Sect. 2.2, in terms of the "Lebesgue measure" λ [B1], we consider concepts of the uniform distribution and the Riemann integrability in infinite-dimensional rectangles in R^∞ and prove infinite-dimensional versions of the famous results of Lebesgue [N1] and Weyl [W], respectively. In this section we show that if $(\alpha_n^{(k)})_{n\in N}$ is an infinite sequence of different integer numbers for every $k \in N$, then a set of all sequences $(x_k)_{k\in N}$ in R^∞ for which a sequence of increasing sets $(Y_n((x_k)_{k\in N}))_{n\in N}$ defined by

$$Y_n((x_k)_{k\in N}) = \prod_{k=1}^{n}((\cup_{j=1}^{n}\{<\alpha_j^{(k)}x_k > (b_k - a_k)\}) + a_k) \times \prod_{k\in N\setminus\{1,\dots,n\}} \{a_k\}$$

is not λ-uniformly distributed on the $\prod_{k\in N}([a_k, b_k])$ is of λ measure zero and, hence shy in R^∞, where $< \cdot >$ denotes the fractional part of the real number.

In Sect. 2.3, a Monte-Carlo algorithm for estimating the value of infinite-dimensional Riemann integrals over infinite-dimensional rectangles in R^∞ described by Pantsulaia [P6] is presented. Further, we introduce Riemann integrability for real-valued functions with respect to product measures in R^∞ and give some sufficient conditions under which a real-valued function of infinitely many real variables is Riemann integrable. We describe a Monte-Carlo algorithm for computing of infinite-dimensional Riemann integrals for such functions.

In Sect. 2.4, we consider some interesting applications of Monte-Carlo algorithms for computing of the infinite-dimensional Riemann integrals described in Sect. 2.3.

Chapter 3 is devoted to study of the structure of the set of all sequences uniformly distributed in $[-1/2, 1/2]$. Pantsulaia [P5] has shown that μ-almost every element of $\mathbf{R}^\infty$ is uniformly distributed in $[-1/2, 1/2]$, where μ denotes Moore-Yamasaki-Kharazishvili measure in $\mathbf{R}^\infty$ for which $\mu([-1/2, 1/2]^\infty) = 1$. In Sect. 3.3 we prove that the set D of all real-valued sequences uniformly distributed in $[-1/2, 1/2]$ is shy in R^N. In Sect. 3.4, we demonstrate that in the Solovay model [So1] the set F of all sequences uniformly distributed modulo 1 in $[-1/2, 1/2]$ is prevalent set [HSY] in R^N.

Chapter 4 contains a brief description of Yamasaki's [Y] remarkable investigation (1980) of the relationship between Moore-Yamasaki-Kharazishvili type measures and infinite powers of Borel diffused probability measures on R. More precisely, Yamasaki's proof is given that no infinite power of the Borel probability measure with a strictly positive density function on R has an equivalent Moore-Yamasaki-Kharazishvili type measure. A certain modification of Yamasaki's example is used for the construction of such a Moore-Yamasaki-Kharazishvili type measure that is equivalent to the product of a certain infinite family of Borel probability measures with a strictly positive density function on R. By virtue of the properties of real-valued sequences equidistributed on the real axis, it is demonstrated that an arbitrary family of infinite powers of Borel diffused probability measures with strictly positive density functions on R is strongly separated and, accordingly, has an infinite-sample well-founded estimator of the unknown distribution function. This extends the main result established in [ZPS].

The last two chapters of the book present applications of the theories of Haar null sets and of uniformly distributed sequences in $[0, 1]$ to statistics.

In Chap. 5, by using the notion of a Haar ambivalent set introduced by Balka, Buczolich and Elekes [BBE], essentially new classes of statistical structures having objective and strong objective estimates of unknown parameters are introduced in a Polish non-locally-compact group admitting an invariant metric and relations between them are studied. An example of a weakly separated statistical structure is constructed for which a question asking "*whether there exists a consistent estimate of an unknown parameter*" is not solvable in the theory (ZF) & (DC). A question asking "*whether there exists an objective consistent estimate of an unknown parameter for any statistical structure in a non-locally compact Polish group with an invariant metric when subjective one exists*" is answered positively in [KKP] when there exists at least one such a parameter the pre-image of which under this subjective estimate is a prevalent set. This construction essentially uses the rather recent celebrated result of Solecki [So2] concerning the partition of a non-locally compact Polish group into a continuous family of pairwise disjoint Haar ambivalents. These results are extensions of recent results of Pantsulaia and Kintsurashvili [PK2]. Some examples of objective and strong objective consistent estimates in a compact Polish group $\{0; 1\}^N$ are also considered in this chapter. At the end of the chapter we present a certain claim for theoretical statisticians in which each consistent estimation with domain in a non-locally compact Polish group equipped with an invariant metric must pass the certification exam on objectivity prior to its practical application and give some recommendations.

In Chap. 6, the notion of Haar null set firstly introduced by Christensen [Ch1] in 1973 and reintroduced in 1992 in the context of dynamical systems by Hunt, Sauer and Yorke [HSY] is used in studying structures of domains of some infinite sample statistics (for example, of an infinite sample average) and in explaining why the null hypothesis is rejected for "almost every" infinite sample by hypothesis testing with maximal reliability.

References

[A] Arbuthnot, J.: An argument for divine providence taken from the constant regularity in the births of both sexes. Philos. Trans. Royal Soc. **27**, 186–190 (1710).

[B1] Baker, R. Lebesgue measure on $\mathbf{R}^\infty$. Proc. Amer. Math. Soc. 113(4) 1023–1029 (1991)

[BBE] Balka, R., Buczolich, Z., Elekes, M.: Topological Hausdorff dimension and level sets of generic continuous functions on fractals. Chaos Solitons Fractals. **45**(12), 1579–1589 (2012)

[BS] Baxa, C., Schoiβengeier, J.: Calculation of improper integrals using $(n\alpha)$-sequences. Monatsh. Math. **135**(4) 265–277 (2002). (Dedicated to Edmund Hlawka on the occasion of his 85th birthday)

[Ch] Chow, S.L.: Statistical Significance: Rationale, Validity and Utility (1997)

[Ch1] Christensen, J.R.: Measure theoretic zero sets in infinite dimensional spaces and applications to differentiability of Lipschitz mappings. Publ. Dep. Math. **10**(2), 29–39 (1973)

[Coh] Cohen, J.: The Earth is round ($p < 0.05$). Am. Psychol. **49**(12), 997–1003 (1994)

[HSY] Hunt, B.R., Sauer, T., Yorke, J.A.: Prevalence: a translation-invariant "Almost Every" on infinite-dimensional spaces. Bull. (New Ser.) Am. Math. Soc. **27**(2), 217–238 (1992)

[K] Kline, R.: Beyond significance testing: reforming data analysis methods in behavioral research, American Psychological Association, Washington, DC (2004)

[KKP] Kintsurashvili, M., Kiria, T., Pantsulaia, G.: On objective and strong objective consistent estimates of unknown parameters for statistical structures in a Polish group admitting an invariant metric. J. Stat. Adv. Theory Appl. **13**(2), 179–233 (2015)

[LMS] Lavoie, H.L., Mulaik, S.A., Steiger, J.H.: What if there were no significance tests? Lawrence Erlbaum Associates (1997)

[MH] Morrison, D., Henkel, R.: The Significance Test Controversy, Aldine Transaction (2006)

[MZ] McCloskey, D.N., Ziliak, S.T.: The Cult of Statistical Significance: How the Standard Error Costs Us Jobs, Justice, and Lives, University of Michigan Press (2008)

[N] Nunnally, J.: The place of statistics in psychology. Educ. Psychol. Measur. **20**(4), 641–650 (1960)

[N1] Nikolski, S.M.: Course of mathematical analysis (in Russian), no. 1, Moscow (1983)

[O] Oakes, M.: Statistical Inference: A Commentary for the Social and Behavioural Sciences. Wiley, Chichester, New York (1986)

[P5] Pantsulaia, G.R.: On uniformly distributed sequences on $[-1/2, 1/2]$. Georg. Inter. J. Sci. Tech. **4**(3), 21–27 (2013)

[P6] Pantsulaia, G., Kiria T.: Calculation of Improper Integrals by Using Uniformly Distributed Sequences. http://arxiv.org/abs/1507.02978

[PK2] Pantsulaia, G., Kintsurashvili, M.: An effective construction of the strong objective infinite sample well-founded estimate. Proc. A. Razmadze Math. Ins. **166**, 113–119 (2014)

[RR] Rodgers, J.L., Rowe, D.C.: Theory development should begin (but not end) with good empirical fits: a comment on Roberts and Pashler (2000). Psychol. Rev. **109**(3) 599–604 (2002)

[RW] Robinson, D.H., Wainer, H.: On the Past and Future of Null Hypothesis Significance Testing. ETC, Statistics and Research Division, Princeton, New York (2001)

[Sol] Solovay, R.M.: A model of set theory in which every set of reals is Lebesgue measurable. Ann. Math. **92**, 1–56 (1970)

[So2] Solecki, S.: On haar null sets. Fund. Math. **149**(3), 205–210 (1996)

[W] Weyl, H.: Úber ein Problem aus dem Gebiete der diophantischen Approximation. Marchr. Ges. Wiss., Gótingen, Math-phys. **Kl**, 234–244 (1916)

[Y] Yamasaki,Y.: Translationally invariant measure on the infinite-dimensional vector space. Publ. Res. Inst. Math. Sci. **16**(3), 693–720 (1980)

[ZPS] Zerakidze, Z., Pantsulaia, G., Saatashvili, G.: On the separation problem for a family of Borel and Baire G-powers of shift-measures on **R**. Ukrainian Math. J. **65**(4), 470–485 (2013)

Chapter 1
Calculation of Improper Integrals by Using Uniformly Distributed Sequences

1.1 Introduction

A useful technique for numerical calculation of the one-dimensional Riemann integral for a real-valued Riemann integrable function over [0, 1] in terms of uniformly distributed sequences was first given in 1916 by Hermann Weyl's celebrated theorem as follows.

Theorem 1.1.1 ([KN], Corollary 1.1, p. 3) *The sequence of real numbers* $(x_n)_{n \in N} \in [0,1]^{\infty}$ *is uniformly distributed in* $[0, 1]$ *if and only if for every real-valued Riemann integrable function* f *on* $[0, 1]$ *we have*

$$\lim_{N \to \infty} \frac{\sum_{n=1}^{N} f(x_n)}{N} = \int_0^1 f(x)dx. \tag{1.1.1}$$

Main corollaries of this theorem were used successfully in Diophantine approximations and have applications to Monte Carlo integration (see, e.g., [KN, H1, H2]). During the last decades the methods of the theory of uniform distribution modulo one have been intensively used for calculation of improper Riemann integrals (see, e.g., [S, BS]).

Note that the set S of all uniformly distributed sequences in $[0, 1]$ viewed as a subset of $[0, 1]^{\infty}$ has full ℓ_1^{∞}-measure, where ℓ_1^{∞} denotes the infinite power of the linear Lebesgue measure ℓ_1 in $[0, 1]$. Therefore each element of the set S can be used for calculation of the one-dimensional Riemann integral for an arbitrary Riemann integrable real-valued function in $[0, 1]$. For an arbitrary Lebesgue integrable function f in $[0, 1]$, there naturally arises the following question.

Question 1.1.1 *What is a maximal subset* S_f *of* S*, each element of which can be used for calculation of the Lebesgue integral over* $[0, 1]$ *by the formula (1.1.1), if this subset has the full* ℓ_1^{∞}*-measure?*

G. Pantsulaia, *Applications of Measure Theory to Statistics*,
DOI 10.1007/978-3-319-45578-5_1

In this chapter we consider two tasks:

The first task is an investigation of Question 1.1.1 by using Kolmogorov's strong law of large numbers.

The second task is an improvement of the following result of C. Baxa and J. Schoiβengeier.

Theorem 1.1.2 ([BS], Theorem 1, p. 271) *Let α be an irrational number, $\mathbf{Q}$ be a set of all rational numbers, and $F \subseteq [0,1] \cap \mathbf{Q}$ be finite. Let $f : [0,1] \to R$ be an integrable, continuous almost everywhere, and locally bounded on $[0,1] \setminus F$. Assume further that for every $\beta \in F$ there is some neighborhood U of β such that f is either bounded or monotone in $[0,\beta) \cap U$ and in $(\beta, 1] \cap U$ as well. Then the following conditions are equivalent.*

1. $\lim_{n\to\infty} \frac{f(x_n)}{n} = 0$.
2. $\lim_{N\to\infty} \frac{1}{N} \sum_{k=1}^{N} f(x_k)$ *exists.*
3. $\lim_{N\to\infty} \frac{1}{N} \sum_{k=1}^{N} f(x_k) = \int_{(0,1)} f(x)dx$.

More precisely, we present an extension of the result of Theorem 1.1.2 to maximal sets $D_f \subset S$ and $E_f \subseteq (0,1)^\infty$ strictly containing all sequences of the form $(\{\alpha n\})_{n\in\mathbf{N}}$ where α is an irrational number and calculate ℓ_1^∞ measures of D_f and E_f, respectively; This Chapter is organized as follows.

In Sect. 1.2 we consider some auxiliary notions and facts from the theory of uniformly distributed sequences on the interval $[0,1]$. In Sect. 1.3 we present proofs of a certain modification of the Kolmogorov strong law of large numbers and the Glivenko–Cantelli theorem. In Sect. 1.4 we consider an application of the Kolmogorov strong law of large numbers to extension of the main result of Baxa and Schoiβengeier [BS].

1.2 Some Auxiliary Notions and Facts from the Theory of Uniform Distribution of Sequences

Definition 1.2.1 A sequence $s_1, s_2, s_3, \ldots$ of real numbers from the interval $[a,b]$ is said to be uniformly distributed in the interval $[a,b]$ if for any subinterval $[c,d]$ of the $[a,b]$ we have

$$\lim_{n\to\infty} \frac{\#(\{s_1, s_2, s_3, \ldots, s_n\} \cap [c,d])}{n} = \frac{d-c}{b-a}, \tag{1.2.1}$$

where # denotes a counting measure.

Definition 1.2.2 The sequence $s_1, s_2, s_3, \ldots$ is said to be uniformly distributed modulo 1 if the sequence $(\{s_n\})_{n\in N}$ of the fractional parts of the $(s_n)_{n\in N}$, is equidistributed (equivalently, uniformly distributed) in the interval $[0,1]$.

Example 1.2.1 ([KN], *Exercise 1.12, p. 16*) The sequence of all multiples of an irrational α

$$0, \alpha, 2\alpha, 3\alpha \dots \tag{1.2.2}$$

is uniformly distributed modulo 1.

Remark 1.2.1 Note that a sequence $(x_k)_{k \in N} \in (0,1)^\infty$ is uniformly distributed in $[0,1]$ if and only if it is equidistributed modulo 1.

The following lemma contains an interesting application of uniformly distributed sequences in $(0,1)$ for a calculation of the Riemann integral over the one-dimensional unit interval $[0,1]$.

Lemma 1.2.1 (Weyl [W]) *These two conditions are equivalent:*
(i) $(a_n)_{n \in N}$ *is equidistributed modulo* 1.
(ii) for every Riemann integrable function f *on* $[0,1]$

$$\lim_{n \to \infty} \frac{1}{n} \sum_{j=1}^{n} f(\{a_j\}) = \int_{[0,1]} f(x)dx. \tag{1.2.3}$$

Lemma 1.2.2 ([N1] Lebesgue Theorem, p. 359) *Let* f *be a bounded real-valued function on* $[0,1]$. *Then* f *is Riemann integrable on* $[0,1]$ *if and only if* f *is* ℓ_1 *almost everywhere continuous on* $[0,1]$.

Lemma 1.2.3 ([KN] Lemma 2.1, p. 182) *Let* $\mathcal{B}[0,1]$ *be a set of all bounded Borel measurable functions on* $[0,1]$ *and let* ℓ_1^∞ *be the infinite power of the standard linear Lebesgue measure* ℓ_1 *on* $[0,1]$. *Then for* $f \in \mathcal{B}[0,1]$, *we have*

$$\ell_1^\infty(\{(x_k)_{k \in \mathbf{N}} : (x_k)_{k \in \mathbf{N}} \in [0,1]^\infty \ \& \ \lim_{N \to \infty} \frac{1}{N} \sum_{n=1}^{N} f(x_n) = \int_{[0,1]} f(x)dx\}) = 1.$$

Proof Note that it is sufficient to prove Lemma 1.2.3 for $f \in \mathcal{B}[0,1]$ with

$$\int_{[0,1]} f(x)dx = 0.$$

Indeed, if $\int_{[0,1]} f(x)dx = c \neq 0$, then we consider a function g defined by $g(x) = f(x) - c$ for $x \in [0,1]$. The validity of Lemma 1.2.3 for g implies

$$\ell_1^\infty(\{(x_k)_{k \in \mathbf{N}} : (x_k)_{k \in \mathbf{N}} \in [0,1]^\infty \ \& \ \lim_{N \to \infty} \frac{1}{N} \sum_{n=1}^{N} (f(x_n) - c) = \int_{[0,1]} (f(x) - c)dx\}) = 1,$$

which is equivalent to the condition

$$\ell_1^\infty(\{(x_k)_{k \in \mathbf{N}} : (x_k)_{k \in \mathbf{N}} \in [0,1]^\infty \ \& \ \lim_{N \to \infty} \frac{1}{N} \sum_{n=1}^{N} f(x_n) = \int_{[0,1]} f(x)dx\}) = 1.$$

For $N \geq 1$, we define F_N on $[0, 1]^\infty$ as follows.

$$F_N(x_1, \ldots, x_n, \ldots) = \frac{1}{N} \sum_{n=1}^{N} f(x_n).$$

Then we get

$$\begin{aligned}
&\int_{[0,1]^\infty} F_N^2(x_1, \ldots) d\ell_1^\infty(x_1, \ldots) \\
&= \frac{1}{N^2} \sum_{n=1}^{N} \int_{[0,1]^\infty} f^2(x_n) d\ell_1^\infty(x_1, \ldots) \\
&+ \frac{2}{N^2} \sum_{1 \leq i < j \leq N} \int_{[0,1]^\infty} f(x_i) f(x_j) d\ell_1^\infty(x_1, \cdots) \\
&= \frac{1}{N^2} \sum_{n=1}^{N} \int_{[0,1]} f^2(x) dx + \frac{2}{N^2} \sum_{1 \leq i < j \leq N} \Big(\int_{[0,1]} f(x) d(x) \Big)^2 \\
&= \frac{1}{N^2} \sum_{n=1}^{N} \int_{[0,1]} f^2(x) dx = \frac{\int_{[0,1]} f^2(x) dx}{N}.
\end{aligned} \tag{1.2.4}$$

Hence

$$\begin{aligned}
&\sum_{m=1}^{\infty} \int_{[0,1]^\infty} F_{m^2}^2(x_1, \ldots) d\ell_1^\infty(x_1, \ldots) \\
&= \int_{[0,1]} f^2(x) dx \sum_{m=1}^{\infty} \frac{1}{m^2} < \infty.
\end{aligned} \tag{1.2.5}$$

By Levi's well-known theorem, we deduce that the sequence $(F_{m^2})_{m \geq 1}$ tends to zero and m tends to $+\infty$ ℓ_1^∞ almost everywhere on $[0, 1]^\infty$. For $N \geq 1$, we can choose $m \geq 1$ such that $m^2 \leq N < (m + 1)^2$. Then we get

$$\begin{aligned}
&|F_N| = \Big|\frac{1}{N}\big(f(x_1) + \cdots + f(x_{m^2})\big) + \frac{1}{N}\big(f(x_{m^2+1}) \\
&+ \cdots + f(x_N)\big)\Big| = \Big|\frac{m^2}{N} \times \frac{1}{m^2}\big(f(x_1) \\
&+ \cdots + f(x_{m^2})\big) + \frac{1}{N}\big(f(x_{m^2+1}) + \cdots + f(x_N)\big)\Big|
\end{aligned} \tag{1.2.6}$$

$$\begin{aligned}
&\leq |F_{m^2}| + \frac{2m}{N} ||f|| = |F_{m^2}| + \frac{2m^2}{mN} ||f|| \\
&\leq |F_{m^2}| + \frac{2}{m} ||f||,
\end{aligned} \tag{1.2.7}$$

where $||f|| = \sup_{x \in [0,1]} f(x)$.

Because the right-hand side of the last equality tends to zero when m tends to $+\infty$ (equivalently, N tends to $+\infty$), we end the proof of Lemma 1.2.3.

Lemma 1.2.4 ([KN] Theorem 2.2, p. 183) *Let S be a set of all elements of* $[0, 1]^\infty$ *that are uniformly distributed on* $[0, 1]$. *Then* $\ell_1^\infty(S) = 1$.

Proof Let $(f_k)_{k \in \mathbf{N}}$ be a countable subclass of $\mathscr{B}[0, 1]$ that defines a uniform convergence on $[0, 1]$.[1] For $k \in N$, we set

$$B_k = \left\{ (x_k)_{k \in \mathbf{N}} : (x_k)_{k \in \mathbf{N}} \in [0, 1]^\infty \ \& \ \lim_{N \to \infty} \frac{1}{N} \sum_{n=1}^{N} f_k(x_n) = \int_{[0,1]} f_k(x)dx \right\}.$$

By Lemma 1.2.3 we know that $\ell_1^\infty(B_k) = 1$ for $k \in \mathbf{N}$, which implies $\ell_1^\infty(\cap_{k \in \mathbf{N}} B_k) = 1$. Thus

$$\ell_1^\infty \Bigg(\Bigg\{ (x_k)_{k \in \mathbf{N}} : (x_k)_{k \in \mathbf{N}} \in [0, 1]^\infty \ \& \ (\forall k)(k \in \mathbf{N} \to \lim_{N \to \infty} \frac{1}{N} \sum_{n=1}^{N} f_k(x_n)$$
$$= \int_{[0,1]} f_k(x)dx \Bigg\} \Bigg) = 1. \tag{1.2.8}$$

The latter relation means that ℓ_1^∞ almost every element of $[0, 1]^\infty$ is uniformly distributed on $[0, 1]$.

1.3 Kolmogorov Strong Law of Large Numbers and Glivenko–Cantelli Theorem

We need some auxiliary facts from mathematical analysis and probability theory.

Lemma 1.3.1 (Kolmogorov–Khinchin ([Sh], Theorem 1, p. 371)) *Let (X, S, μ) be a probability space and let $(\xi_n)_{n \in \mathbf{N}}$ be the sequence of independent random variables for which $\int_X \xi_n(x)d\mu(x) = 0$. If $\sum_{n=1}^{\infty} \int_X \xi_n^2(x)d\mu(x) < \infty$, then the series $\sum_{n=1}^{\infty} \xi_n$ converges with probability* 1.

Proof We put $S_n = \sum_{k=1}^{n} \xi_k$ for $n \in \mathbf{N}$. We set

$$A = \{x : x \in X \ \& \ \max |S_k(x)| \geq \varepsilon\},$$

[1] We say that a family $(f_k)_{k \in \mathbf{N}}$ of elements of $\mathscr{B}[0, 1]$ defines a uniform convergence on $[0, 1]$, if for each $(x_n)_{n \in N} \in [0, 1]^\infty$ the validity of the condition $\lim_{N \to \infty} \frac{1}{N} \sum_{n=1}^{N} f_k(x_n) = \int_{[0,1]} f_k(x)dx$ for $k \in N$ implies that $(x_n)_{n \in N}$ is uniformly distributed on $[0, 1]$. Indicator functions of closed subintervals of $[0, 1]$ with rational endpoints is an example of such a family.

$$A_k = \{x : x \in X \ \& \ |S_i(x)| < \varepsilon \text{ for } i = 1, \ldots, k-1 \ \& \ |S_k(x)| \geq \varepsilon\}$$

for $k = 1, 2, \ldots, n$.

It is obvious that $A = \sum_{k=1}^{n} A_k$ and

$$\int_X S_n^2(x) d\mu(x) \geq \int_X S_n^2(x) \times \chi_A(x) d\mu(x) = \sum_{k=1}^{n} \int_X S_n^2(x) \times \chi_{A_k}(x) d\mu(x).$$

Because $\sum_{i=1}^{k} \xi_i \times \chi_{A_k}$ and $\sum_{i=k+1}^{n} \xi_i$ are independent random variables and $\int_X \xi_k(x) d\mu(x) = 0$ for $k = 1, \ldots, n$, we get

$$\int_X S_k(x) \times \sum_{i=k+1}^{n} \xi_i(x) \times \chi_{A_k}(x) d\mu(x)$$
$$= \int_X S_k(x) \times \chi_{A_k}(x) d\mu(x) \times \int_X \sum_{i=k+1}^{n} \xi_i(x) d\mu(x) = 0. \quad (1.3.1)$$

The latter relation implies

$$\int_X S_n^2(x) \times \chi_{A_k}(x) d\mu(x) = \int_X (S_k(x) + \sum_{i=k+1}^{n} \xi_i(x))^2 \times \chi_{A_k}(x) d\mu(x)$$
$$= \int_X S_k^2(x) \times \chi_{A_k}(x) d\mu(x) + 2 \int_X S_k(x) \times \sum_{i=k+1}^{n} \xi_i(x) \times \chi_{A_k}(x) d\mu(x)$$
$$+ \int_X \left(\sum_{i=k+1}^{n} \xi_i(x) \right)^2 \times \chi_{A_k}(x) d\mu(x) \geq \int_X S_k^2(x) \times \chi_{A_k}(x) d\mu(x), \quad (1.3.2)$$

which means that

$$\int_X S_n^2(x) d\mu(x) \geq \int_X S_n^2(x) \times \chi_A(x) d\mu(x) = \sum_{k=1}^{n} \int_X S_n^2(x) \times \chi_{A_k}(x) d\mu(x)$$
$$\geq \sum_{k=1}^{n} \int_X S_k^2(x) \times \chi_{A_k}(x) d\mu(x) \geq \varepsilon^2 \times \sum_{k=1}^{n} \mu(A_k) = \varepsilon^2 \times \mu(A). \quad (1.3.3)$$

Thus we have obtained the validity of the inequality

$$\mu\{x : x \in X \ \& \ \max_{1 \leq k \leq n} |S_k(x)| \geq \varepsilon\} \leq \frac{1}{\varepsilon^2} \int_X S_n^2(x) d\mu(x).$$

It is well known that the sequence $(S_n)_{n \in \mathbf{N}}$ converges for μ almost every point if and only if this sequence is fundamental for μ almost every point. But we know that $(S_n)_{n\in\mathbf{N}}$ is fundamental for μ almost every point if and only if the following condition

$$\lim_{n\to\infty}\mu\{x : \sup_{k\geq 1}|S_{n+k}(x) - S_k(x)| \geq \varepsilon\} = 0$$

holds true. We get

$$\mu\{x : \sup_{k\geq 1}|S_{n+k}(x) - S_k(x)| \geq \varepsilon\} = \lim_{N\to\infty}\mu\{x : x \in X \ \&\ \max_{1\leq k\leq N}|S_{n+k}(x) - S(n)| \geq \varepsilon\}$$

$$\leq \lim_{N\to\infty}\frac{1}{\varepsilon^2}\sum_{k=n}^{n+N}\int_X \xi_k^2(x)d\mu(x) = \frac{1}{\varepsilon^2}\sum_{k=n}^{\infty}\int_X \xi_k^2(x)d\mu(x). \tag{1.3.4}$$

Finally we get

$$\lim_{n\to\infty}\mu\{x : \sup_{k\geq 1}|S_{n+k}(x) - S_k(x)| \geq \varepsilon\} \leq \lim_{n\to\infty}\frac{1}{\varepsilon^2}\sum_{k=n}^{\infty}\int_X \xi_k^2(x)d\mu(x) = 0.$$

This ends the proof of Lemma 1.3.1.

Lemma 1.3.2 (Toeplitz Lemma ([Sh], Lemma 1, p. 377)) *Let $(a_n)_{n\in\mathbf{N}}$ be a sequence of nonnegative numbers, $b_n = \sum_{i=1}^n a_i$, $b_n > 0$ for each $n \geq 1$ and $b_n \uparrow \infty$, when $n \to \infty$. Let $(x_n)_{n\in\mathbf{N}}$ be a sequence of real numbers such that* $\lim_{n\to\infty} x_n = x$. *Then*

$$\lim_{n\to\infty}\frac{1}{b_n}\sum_{j=1}^{n} a_j x_j = x.$$

In particular, if $a_n = 1$ for $n \in \mathbf{N}$, then

$$\lim_{n\to\infty}\frac{1}{n}\sum_{k=1}^{n} x_k = x.$$

Proof For $\varepsilon > 0$, let $n_0 = n_0(\varepsilon)$ be a natural number such that $|x_n - x| < \frac{\varepsilon}{2}$ for $n \geq n_0$. Let us choose $n_1 > n_0$ such that

$$\frac{1}{b_{n_1}}\sum_{j=1}^{n_0} a_j|x_j - x| < \frac{\varepsilon}{2}.$$

Then for $n > n_1$, we get

$$|\frac{1}{b_n}\sum_{j=1}^{n} a_j x_j - x| \leq \frac{1}{b_n}\sum_{j=1}^{n} a_j|x_j - x|$$

$$= \frac{1}{b_n}\sum_{j=1}^{n_0} a_j|x_j - x| + \frac{1}{b_n}\sum_{j=n_0+1}^{n} a_j|x_j - x|$$

$$\leq \frac{1}{b_{n_1}}\sum_{j=1}^{n_0} a_j|x_j - x| + \frac{1}{b_n}\sum_{j=n_0+1}^{n} a_j|x_j - x|$$

$$\leq \frac{\varepsilon}{2} + \frac{\sum_{j=n_0+1}^{n} a_j}{\sum_{j=0}^{n} a_j}\frac{\varepsilon}{2} \leq \varepsilon. \tag{1.3.5}$$

Lemma 1.3.3 (Kronecker Lemma ([Sh], Lemma 2, p. 378)) *Let $(b_n)_{n\in\mathbf{N}}$ be an increasing sequence of positive numbers such that $b_n \uparrow \infty$, when $n \to \infty$, and let $(x_n)_{n\in\mathbf{N}}$ be a sequence of real numbers such that the series $\sum_{k\in\mathbf{N}} x_k$ converges. Then*

$$\lim_{n\to\infty}\frac{1}{b_n}\sum_{j=1}^{n} b_j x_j = 0.$$

In particular, if $b_n = 0$, $x_n = \frac{y_n}{n}$ and the series $\sum_{n=1}^{\infty}\frac{y_n}{n}$ converges then

$$\lim_{n\to\infty}\frac{\sum_{k=1}^{n} y_k}{n} = 0.$$

Proof Let $b_0 = 0$, $s_0 = 0$, $S_n = \sum_{j=1}^{n} x_j$. Then we get

$$\sum_{j=1}^{n} b_j x_j = \sum_{j=1}^{n} b_j(S_j - S_{j-1}) = b_n S_n - b_0 S_0 - \sum_{j=1}^{n} S_{j-1}(b_j - b_{j-1}).$$

Because $\lim_{n\to\infty} S_n = x$ for some $x \in R$, by the Toeplitz lemma (cf. Lemma 1.3.2) we get that

$$\lim_{n\to\infty}\frac{1}{b_n}\sum_{j=1}^{n} S_{j-1}a_j = x,$$

where $a_j = b_j - b_{j-1}$ for $1 \leq j \leq n$. By using the latter relation we easily conclude that

$$\lim_{n\to\infty}\frac{1}{b_n}\sum_{j=1}^{n} b_j x_j = \lim_{n\to\infty} S_n - \lim_{n\to\infty}\frac{1}{b_n}\sum_{j=1}^{n} S_{j-1}a_j = 0,$$

which implies

$$\lim_{n\to\infty}\frac{1}{b_n}\sum_{j=1}^{n} b_j x_j = 0.$$

Below we give the proof of a certain modification of the Kolmogorov strong law of large numbers (cf. [Sh], Theorem 3, p. 379).

Theorem 1.3.1 *Let $(X, \mathbf{F}, \mu)$ be a probability space and let $\mathbf{L}(X)$ be a class of all real-valued Lebesgue measurable functions on X. Let μ^∞ be the infinite power of the probability measure μ. Then for $f \in \mathbf{L}(X)$ we have $\mu^\infty(A_f) = 1$, where A_f is defined by*

$$A_f = \left\{ (x_k)_{k\in\mathbf{N}} : (x_k)_{k\in\mathbf{N}} \in X^\infty \ \& \ \lim_{N\to\infty} \frac{1}{N} \sum_{n=1}^{N} f(x_n) = \int_X f(x)d\mu(x) \right\}. \tag{1.3.6}$$

Proof Without loss of generality, we can assume that f is nonnegative. We put $\xi_k((x_i)_{i\in\mathbf{N}}) = f(x_k)$ for $k \in \mathbf{N}$ and $(x_i)_{i\in\mathbf{N}} \in X^\infty$. We also put

$$\eta_k((x_i)_{i\in\mathbf{N}}) = \frac{1}{k}\left[\xi_k((x_i)_{i\in\mathbf{N}})\chi_{\{\omega:\xi_k(\omega)<k\}}((x_i)_{i\in\mathbf{N}}) - \int_{X^\infty} \xi_k((z_i)_{i\in\mathbf{N}})\chi_{\{\omega:\xi_k(\omega)<k\}}((z_i)_{i\in\mathbf{N}})d\mu^\infty((z_i)_{i\in\mathbf{N}})\right] \tag{1.3.7}$$

for $(x_i)_{i\in\mathbf{N}} \in X^\infty$.

Note that $(\eta_k)_{k\in\mathbf{N}}$ is the sequence of independent random variables for which $\int_{X^\infty} \eta_k d\mu^\infty = 0$.

We have

$$\sum_{n=1}^{\infty} \int_{X^\infty} \eta_n^2((x_i)_{i\in\mathbf{N}})d\mu^\infty((x_i)_{i\in\mathbf{N}})$$
$$= \sum_{n=1}^{\infty} \frac{1}{n^2} \int_{X^\infty} \xi_n^2((x_i)_{i\in\mathbf{N}})\chi_{\{(y_i)_{i\in\mathbf{N}}:\xi_n((y_i)_{i\in\mathbf{N}})<n\}}((x_i)_{i\in\mathbf{N}})d\mu^\infty((x_i)_{i\in\mathbf{N}})$$
$$- \sum_{n=1}^{\infty} \frac{1}{n^2} \left(\int_{X^\infty} \xi_n((x_i)_{i\in\mathbf{N}})\chi_{\{(y_i)_{i\in\mathbf{N}}:\xi_n((y_i)_{i\in\mathbf{N}})<n\}}((x_i)_{i\in\mathbf{N}})d\mu^\infty((x_i)_{i\in\mathbf{N}})\right)^2$$
$$= \sum_{n=1}^{\infty} \frac{1}{n^2} \int_{X^\infty} f(x_n)^2\chi_{\{(y_i)_{i\in\mathbf{N}}:f(y_n)<n\}}((x_i)_{i\in\mathbf{N}})d\mu^\infty((x_i)_{i\in\mathbf{N}})$$
$$- \sum_{n=1}^{\infty} \frac{1}{n^2} \left(\int_{X^\infty} f(x_n)\chi_{\{(y_i)_{i\in\mathbf{N}}:f(y_n)<n\}}((x_i)_{i\in\mathbf{N}})d\mu^\infty((x_i)_{i\in\mathbf{N}})\right)^2$$
$$= \sum_{n=1}^{\infty} \frac{1}{n^2} \int_X f^2(x)\chi_{\{\omega:f(\omega)<n\}}(x)d\mu(x) - \sum_{n=1}^{\infty} \frac{1}{n^2} \left(\int_X f(x)\chi_{\{\omega:f(\omega)<n\}}(x)d\mu(x)\right)^2$$
$$\leq \sum_{n=1}^{\infty} \frac{1}{n^2} \int_X f^2(x)\chi_{\{\omega:f(\omega)<n\}}(x)d\mu(x) = \sum_{n=1}^{\infty} \frac{1}{n^2} \sum_{k=1}^{n} \int_X f^2(x)\chi_{\{\omega:k-1\leq f(\omega)<k\}}(x)d\mu(x)$$
$$= \sum_{k=1}^{\infty} \int_X f^2(x)\chi_{\{\omega:k-1\leq f(\omega)<k\}}(x)d\mu((x)) \sum_{n=k}^{\infty} \frac{1}{n^2} \leq 2 \sum_{k=1}^{\infty} \frac{1}{k} \int_X f^2(x)\chi_{\{\omega:k-1\leq f(\omega)<k\}}(x)d\mu(x)$$
$$\leq 2 \sum_{k=1}^{\infty} \int_X f(x)\chi_{\{\omega:k-1\leq f(\omega)<k\}}(x)d\mu((x)) = 2\int_X f(x)d\mu(x). \tag{1.3.8}$$

Because

$$\sum_{n=1}^{\infty}\int_X \eta_n^2((x_i)_{i\in\mathbf{N}})d\mu((x_i)_{i\in\mathbf{N}}) < +\infty, \tag{1.3.9}$$

by using Lemma 1.3.1 we get

$$\mu\Bigg\{(x_i)_{i\in\mathbf{N}} : \sum_{k=1}^{\infty}\frac{1}{k}\Bigg[f(x_k)\chi_{\{(y_i)_{i\in\mathbf{N}}:f(y_k)<k\}}((x_i)_{i\in\mathbf{N}})$$
$$-\int_{X^\infty}\xi_k((z_i)_{i\in\mathbf{N}})\chi_{\{(y_i)_{i\in\mathbf{N}}:f(y_k)<k\}}((z_i)_{i\in\mathbf{N}})d\mu^\infty((z_i)_{i\in\mathbf{N}})\Bigg] \text{ is convergent}\Bigg\} = 1. \tag{1.3.10}$$

Now by the Kronecker lemma (cf. Lemma 1.3.3) we get that

$$\mu^\infty\Bigg\{(x_i)_{i\in\mathbf{N}} : \lim_{N\to\infty}\frac{1}{N}\sum_{k=1}^{N}\Bigg[f(x_k)\chi_{\{(y_i)_{i\in\mathbf{N}}:f(y_k)<k\}}((x_i)_{i\in\mathbf{N}})$$
$$-\int_{X^\infty}\xi_k((z_i)_{i\in\mathbf{N}})\chi_{\{(y_i)_{i\in\mathbf{N}}:f(y_k)<k\}}((z_i)_{i\in\mathbf{N}})d\mu^\infty((z_i)_{i\in\mathbf{N}})\Bigg] = 0\Bigg\} = 1. \tag{1.3.11}$$

Note that

$$\sum_{n=1}^{\infty}\mu^\infty(\{(x_i)_{i\in\mathbf{N}} : \xi_1((x_i)_{i\in\mathbf{N}}) \geq n\})$$
$$= \sum_{n=1}^{\infty}\sum_{k\geq n}\mu^\infty\{(x_i)_{i\in\mathbf{N}} : k \leq \xi_1((x_i)_{i\in\mathbf{N}}) < k+1\}$$
$$= \sum_{k=1}^{\infty}k\mu^\infty\{(x_i)_{i\in\mathbf{N}} : k \leq \xi_1((x_i)_{i\in\mathbf{N}}) < k+1\}$$
$$= \sum_{k=0}^{\infty}\int_{X^\infty}k\chi_{\{(y_j)_{j\in\mathbf{N}}:k\leq\xi_1((y_i)_{i\in\mathbf{N}})<k+1\}}((z_i)_{i\in\mathbf{N}})d\mu^\infty((z_i)_{i\in\mathbf{N}})$$
$$\leq \sum_{k=0}^{\infty}\int_{X^\infty}\xi_1((z_i)_{i\in\mathbf{N}})\chi_{\{(y_j)_{j\in\mathbf{N}}:k\leq\xi_1((y_j)_{j\in\mathbf{N}})<k+1\}}((z_i)_{i\in\mathbf{N}})d\mu^\infty((z_i)_{i\in\mathbf{N}})$$
$$= \int_{X^\infty}\xi_1((z_i)_{i\in\mathbf{N}})d\mu^\infty((z_i)_{i\in\mathbf{N}}) < +\infty. \tag{1.3.12}$$

Because $(\xi_k)_{k\in\mathbf{N}}$ is a sequence of equally distributed random variables on X^∞, we have

$$\sum_{n=1}^{\infty}\mu^{\infty}(\{(x_i)_{i\in\mathbf{N}} : \xi_k((x_i)_{i\in\mathbf{N}}) \geq n\}) \leq \int_{X^{\infty}} \xi_1((x_i)_{i\in\mathbf{N}})d\mu^{\infty}((x_i)_{i\in\mathbf{N}}) < +\infty, \tag{1.3.13}$$

which by the well-known Borelli–Cantelli lemma implies that

$$\mu^{\infty}(\{(x_i)_{i\in\mathbf{N}} : \xi_n((x_i)_{i\in\mathbf{N}}) \geq n\}\ i.o.) = 0. \tag{1.3.14}$$

The latter relation means that

$$\mu^{\infty}(\{(x_i)_{i\in\mathbf{N}} : (\exists N((x_i)_{i\in\mathbf{N}}))(\forall n \geq N((x_i)_{i\in\mathbf{N}}) \to \xi_n((x_i)_{i\in\mathbf{N}}) < n\}) = 1. \tag{1.3.15}$$

Thus, we have obtained the validity of the equality $\mu^{\infty}(A^*_f) = 1$, where

$$A^*_f = \Bigg\{(x_i)_{i\in\mathbf{N}} : \lim_{N\to\infty}\frac{1}{N}\sum_{k=1}^{N}\Bigg[f(x_k)\chi_{\{(y_i)_{i\in\mathbf{N}}:f(y_k)<k\}}((x_i)_{i\in\mathbf{N}})$$
$$-\int_{X^{\infty}}\xi_k((z_i)_{i\in\mathbf{N}})\chi_{\{(y_i)_{i\in\mathbf{N}}:f(y_k)<k\}}((z_i)_{i\in\mathbf{N}})d\mu^{\infty}((z_i)_{i\in\mathbf{N}})\Bigg] = 0$$
$$\&(\exists N((x_i)_{i\in\mathbf{N}}))(\forall n > N((x_i)_{i\in\mathbf{N}}) \to \xi_n((x_i)_{i\in\mathbf{N}}) < n)\Bigg\}. \tag{1.3.16}$$

Now it is obvious that for $(x_i)_{i\in\mathbf{N}} \in A^*_f$, we have

$$0 = \lim_{N\to\infty}\frac{1}{N}\sum_{k=1}^{N}\Bigg[f(x_k)\chi_{\{(y_i)_{i\in\mathbf{N}}:f(y_k)<k\}}((x_i)_{i\in\mathbf{N}})$$
$$-\int_{X^{\infty}}\xi_k((z_i)_{i\in\mathbf{N}})\chi_{\{(y_i)_{i\in\mathbf{N}}:f(y_k)<k\}}((z_i)_{i\in\mathbf{N}})d\mu^{\infty}((z_i)_{i\in\mathbf{N}})\Bigg]$$
$$= \lim_{N\to\infty}\frac{1}{N}\sum_{k=N((x_i)_{i\in\mathbf{N}})}^{N}\Bigg[f(x_k)\chi_{\{(y_i)_{i\in\mathbf{N}}:f(y_k)<k\}}((x_i)_{i\in\mathbf{N}})$$
$$-\int_{X^{\infty}}\xi_k((z_i)_{i\in\mathbf{N}})\chi_{\{(y_i)_{i\in\mathbf{N}}:f(y_k)<k\}}((z_i)_{i\in\mathbf{N}})d\mu^{\infty}((z_i)_{i\in\mathbf{N}})\Bigg]$$
$$= \lim_{N\to\infty}\frac{1}{N}\sum_{k=N((x_i)_{i\in\mathbf{N}})}^{N}\Bigg[f(x_k) - \int_X f(x)\chi_{\{y:f(y)<k\}}(x)d\mu(x)\Bigg]$$
$$= \lim_{N\to\infty}\frac{1}{N}\sum_{k=1}^{N}\Bigg[f(x_k) - \int_X f(x)\chi_{\{y:f(y)<k\}}(x)d\mu(x)\Bigg]. \tag{1.3.17}$$

Inasmuch as

$$\lim_{k\to\infty}\int_X f(x)\chi_{\{y:f(y)<k\}}(x)d\mu(x) = \int_X f(x)d\mu(x), \tag{1.3.18}$$

by Toeplitz's lemma (cf. Lemma 1.3.2) we get

$$\lim_{N\to\infty}\frac{1}{N}\sum_{k=1}^{N}\int_X f(x)\chi_{\{y:f(y)<k\}}(x)d\mu(x) = \int_X f(x)d\mu(x) \tag{1.3.19}$$

which implies that

$$\lim_{N\to\infty}\frac{1}{N}\sum_{k=1}^{N} f(x_k) = \int_X f(x)d\mu(x) \tag{1.3.20}$$

for each $(x_i)_{i\in\mathbf{N}} \in A_f^*$.

The validity of the inclusion $A_f^* \subseteq A_f$ ends the proof of Theorem 1.3.1.

Remark 1.3.1 (*Kolmogorov Strong Law of Large Numbers,* [Sh], *Theorem 3, p. 379*) Kolmogorov's strong law of large numbers states that if $(\Omega, \mathscr{S}, P)$ is a probability space and $(\xi_k)_{k\in\mathbf{N}}$ is a sequence of independent equally distributed random variables for which mathematical expectation m of ξ_1 is finite then the following condition

$$P\left(\left\{\omega : \omega \in \Omega \ \& \ \lim_{n\to\infty}\frac{\sum_{k=1}^{n}\xi_k(\omega)}{n} = m\right\}\right) = 1$$

holds true.

Note the validity of Theorem 1.3.1 can be obtained by the Kolmogorov strong law of large numbers if we put $(\Omega, \mathscr{S}, P) = (X^\infty, \mathbf{F}^\infty, \mu^\infty)$ and $\xi_k((x_i)_{i\in N}) = f(x_k)$ for each $(x_i)_{i\in N} \in X^\infty$, where $(X, \mathbf{F}, \mu)$ come from Theorem 1.3.1.

Theorem 1.3.2 (Glivenko–Cantelli Theorem) *Let F be a distribution function and let P_F be a Borel probability measure in R with distribution function F. Then*

$$P_F^\infty(\{(x_k)_{k\in N} : (x_k)_{k\in N} \in R^\infty \ \& \ \lim_{n\to\infty}\sup_{x\in R}|\frac{\#(\{x_1,\dots,x_n\}\cap(-\infty,x])}{n} - F(x)| = 0\}) = 1.$$

Proof Let us consider a probability space $(\Omega, \mathscr{S}, P) = (R^\infty, \mathscr{B}(R^\infty), P_F^\infty)$. For $x \in R$, we put $\xi_k((x_i)_{i\in N}) = I_{(-\infty,x]}(x_k)$ for $k \in N$. It is obvious that $(\xi_k)_{k\in N}$ is a sequence of equally distributed independent random variables with mathematical expectation $F(x)$. Indeed,

$$\int_{R^\infty}\xi_k((x_i)_{i\in N})dP_F^\infty((x_i)_{i\in N}) = \int_R I_{(-\infty,x]}(x_k)dP_F(x_k) = P_F((-\infty,x]) = F(x).$$

By the strong law of large numbers we get

$$P_F^\infty(\{(x_k)_{k\in N} : (x_k)_{k\in N} \in R^\infty \ \& \ \lim_{n\to\infty}\frac{\sum_{k=1}^{n}\xi_k((x_i)_{i\in N})}{n} = F(x)\}) = 1.$$

But note that

$$\frac{\sum_{k=1}^{n} \xi_k((x_i)_{i \in N})}{n} = \frac{\#(\{x_1, \dots, x_n\} \cap (-\infty, x])}{n}.$$

Thus for $x \in R$, we get

$$P_F^{\infty}(\{(x_k)_{k \in N} : (x_k)_{k \in N} \in R^{\infty} \ \& \ \lim_{n \to \infty} \frac{\#(\{x_1, \dots, x_n\} \cap (-\infty, x])}{n} = F(x)\}) = 1.$$

Similarly, for $x \in R$, we can prove that

$$P_F^{\infty}(\{(x_k)_{k \in N} : (x_k)_{k \in N} \in R^{\infty} \ \& \ \lim_{n \to \infty} \frac{\#(\{x_1, \dots, x_n\} \cap (-\infty, x))}{n} = F(x^-)\}) = 1,$$

where $F(x^-) = P_F((-\infty, x]) - P_F(\{x\})$ is the left-hand limit of the function F at point x.

For $\varepsilon > 0$, let $k > 1/\varepsilon$. We consider "knot" points $\kappa_0, \dots, \kappa_k$ for which

$$-\infty = \kappa_0 < \kappa_1 \le \kappa_2 \le \cdots \le \kappa_{k-1} < \kappa_k = \infty$$

and $\kappa_0, \dots, \kappa_k$ defines a partition of R into disjoint intervals such that

$$F(\kappa_j^-) \le \frac{j}{k} \le F(\kappa_j)$$

for $j = 1, \dots, k-1$. Then by the construction, if $\kappa_{j-1} < \kappa_j$ we get

$$F(\kappa_j^-) - F(\kappa_{j-1}) \le \frac{j}{k} - \frac{(j-1)}{k} = \frac{1}{k} < \varepsilon.$$

By the remark above we have

$$P_F^{\infty}(\{(x_k)_{k \in N} : (x_k)_{k \in N} \in R^{\infty} \ \& \ \lim_{n \to \infty} \frac{\#(\{x_1, \dots, x_n\} \cap (-\infty, \kappa_j])}{n} = F(\kappa_j)\}) = 1,$$

and

$$P_F^{\infty}(\{(x_k)_{k \in N} : (x_k)_{k \in N} \in R^{\infty} \ \& \ \lim_{n \to \infty} \frac{\#(\{x_1, \dots, x_n\} \cap (-\infty, \kappa_j))}{n} = F(\kappa_j^-)\}) = 1.$$

Then it immediately follows that, for each j,

$$P_F^{\infty}(\{(x_k)_{k \in N} : (x_k)_{k \in N} \in R^{\infty} \ \& \ \lim_{n \to \infty} |\frac{\#(\{x_1, \dots, x_n\} \cap (-\infty, \kappa_j])}{n} - F(\kappa_j)| = 0\}) = 1,$$

and

$$P_F^{\infty}(\{(x_k)_{k\in N} : (x_k)_{k\in N} \in R^{\infty} \ \& \ \lim_{n\to\infty} |\frac{\#(\{x_1,\dots,x_n\}\cap(-\infty,\kappa_j))}{n} - F(\kappa_j^-)| = 0\}) = 1.$$

The latter relation implies that

$$P_F^{\infty}(\{(x_k)_{k\in N} : (x_k)_{k\in N} \in R^{\infty} \ \& \ \lim_{n\to\infty} \Delta_n((x_k)_{k\in N}) = 0\}) = 1,$$

where

$$\Delta_n((x_k)_{k\in N})$$
$$= \max_{1\le j\le k-1}\{|\frac{\#(\{x_1,\dots,x_n\}\cap(-\infty,\kappa_j])}{n} - F(\kappa_j)|, |\frac{\#(\{x_1,\dots,x_n\}\cap(-\infty,\kappa_j))}{n} - F(\kappa_j^-)|\}. \tag{1.3.21}$$

For any x find the interval within which x lies; that is, identify j such that

$$\kappa_{j-1} \le x < \kappa_j.$$

Then we get

$$\frac{\#(\{x_1,\dots,x_n\}\cap(-\infty,x])}{n} - F(x) \le \frac{\#\{x_1,\dots,x_n\}\cap(-\infty,\kappa_j))}{n} - F(\kappa_{j-1})$$
$$\le \frac{\#(\{x_1,\dots,x_n\}\cap(-\infty,\kappa_j))}{n} - F(\kappa_j^-) + \varepsilon \tag{1.3.22}$$

and

$$\frac{\#(\{x_1,\dots,x_n\}\cap(-\infty,x])}{n} - F(x) \ge \frac{\#(\{x_1,\dots,x_n\}\cap(-\infty,\kappa_{j-1}])}{n} - F(\kappa_j^-)$$
$$\ge \frac{\#(\{x_1,\dots,x_n\}\cap(-\infty,\kappa_{j-1}])}{n} - F(\kappa_{j-1}) - \varepsilon. \tag{1.3.23}$$

Thus we get

$$\frac{\#(\{x_1,\dots,x_n\}\cap(-\infty,\kappa_{j-1}])}{n} - F(\kappa_{j-1}) - \varepsilon \le \frac{\#(\{x_1,\dots,x_n\}\cap(-\infty,x])}{n} - F(x)$$
$$\le \frac{\#(\{x_1,\dots,x_n\}\cap(-\infty,\kappa_j))}{n} - F(\kappa_j^-) + \varepsilon. \tag{1.3.24}$$

The latter relations, for each $x \in R$ give that

$$P_F^{\infty}(\{(x_k)_{k\in N} : (\forall n)(n \in N \Rightarrow |\frac{\#(\{x_1,\dots,x_n\}\cap(-\infty,x])}{n} - F(x)| \le \Delta_n((x_k)_{k\in N}) + \varepsilon)\}) = 1,$$

which implies

$$P_F^{\infty}(\{(x_k)_{k\in N} : (\forall n)(n \in N \Rightarrow \sup_{x\in R} |\frac{\#(\{x_1,\dots,x_n\}\cap(-\infty,x])}{n} - F(x)| \le \Delta_n((x_k)_{k\in N}) + \varepsilon)\}) = 1.$$

Because

$$\{(x_k)_{k\in N}:\ (\forall n)(n\in N\Rightarrow \sup_{x\in R}|\frac{\#(\{x_1,\dots,x_n\}\cap(-\infty,x])}{n}-F(x)|\leq \Delta_n((x_k)_{k\in N})+\varepsilon)\}$$

$$\subseteq \{(x_k)_{k\in N}:\ \lim_{n\to\infty}\sup_{x\in R}|\frac{\#(\{x_1,\dots,x_n\}\cap(-\infty,x])}{n}-F(x)|=\varepsilon)\} \quad (1.3.25)$$

we get

$$P_F^{\infty}(\{(x_k)_{k\in N}:\ \lim_{n\to\infty}\sup_{x\in R}|\frac{\#(\{x_1,\dots,x_n\}\cap(-\infty,x])}{n}-F(x)|=\varepsilon\})=1.$$

Setting

$$A_\varepsilon=\{(x_k)_{k\in N}:\ \lim_{n\to\infty}\sup_{x\in R}|\frac{\#(\{x_1,\dots,x_n\}\cap(-\infty,x])}{n}-F(x)|=\varepsilon\},$$

we get $P_F^{\infty}(A_\varepsilon)=1$ for each $\varepsilon>0$. Now if we put $A=\cap_{\varepsilon>0}A_\varepsilon$, then we obtain

$$P_F^{\infty}(A)=P_F^{\infty}(\cap_{\varepsilon>0}A_\varepsilon)=P_F^{\infty}(\cap_{k=1}^{\infty}A_{1/k})=\lim_{k\to\infty}P_F^{\infty}(A_{1/k})=1,$$

which means that

$$P_F^{\infty}(\{(x_k)_{k\in N}:(x_k)_{k\in N}\in R^{\infty}\ \&\ \lim_{n\to\infty}\sup_{x\in R}|\frac{\#(\{x_1,\dots,x_n\}\cap(-\infty,x])}{n}-F(x)|=0\})=1.$$

This ends the proof of Theorem 1.3.2.

1.4 Calculation of a Certain Improper One-Dimensional Riemann Integral by Using Uniformly Distributed Sequences

By using Theorem 1.3.1 we get the validity of the following assertion.

Theorem 1.4.1 *Let f be a Lebesgue integrable real-valued function on* (0, 1). *Then we have*

$$\ell_1^{\infty}(\{(x_k)_{k\in\mathbf{N}}:(x_k)_{k\in\mathbf{N}}\in[0,1]^{\infty}\ \&(x_k)_{k\in\mathbf{N}}\ \text{is uniformly distributed in}\ (0,1)$$

$$\&\ \lim_{N\to\infty}\frac{1}{N}\sum_{k=1}^{N}f(x_k)=\int_0^1 f(x)dx\})=1. \quad (1.4.1)$$

Proof Note that

$$\{(x_k)_{k\in\mathbf{N}} : (x_k)_{k\in\mathbf{N}} \in [0,1]^\infty$$
$$\&(x_k)_{k\in\mathbf{N}} \text{ is uniformly distributed in } (0,1) \ \& \ \lim_{N\to\infty}\frac{1}{N}\sum_{k=1}^{N} f(x_k) = \int_0^1 f(x)dx\} = S\cap A_f, \tag{1.4.2}$$

where S comes from Lemma 1.2.4 and A_f comes from Theorem 1.3.1 when $(X, \mathbf{F}, \mu) = ((0,1), \mathbf{B}(0,1), \ell_1)$.

Note the answer to Question 1.1.1 contained in the following statement.

Theorem 1.4.2 *The set $S_f = A_f \cap S$ is a maximal subset of S, each element of which can be used for calculation of the Lebesgue integral over* $[0, 1]$ *by the formula (1.1.1) and $\ell_1^\infty(S_f) = 1$.*

Observation 1.4.1 *Let $f : (0,1) \to \mathbf{R}$ be a Lebesgue integrable function. Then we have $A_f \subseteq B_f$, where*

$$B_f = \{(x_k)_{k\in\mathbf{N}} : (x_k)_{k\in\mathbf{N}} \in (0,1)^\infty \ \& \ \lim_{N\to\infty}\frac{1}{N}\sum_{k=1}^{N} f(x_k) \text{ exists}\}. \tag{1.4.3}$$

Observation 1.4.2 *Let $f : (0,1) \to \mathbf{R}$ be a Lebesgue integrable function. Then we have $B_f \subseteq C_f$, where*

$$C_f = \{(x_k)_{k\in\mathbf{N}} : (x_k)_{k\in\mathbf{N}} \in (0,1)^\infty \ \& \ \lim_{N\to\infty}\frac{f(x_N)}{N} = 0\}. \tag{1.4.4}$$

Proof For $(x_k)_{k\in\mathbf{N}} \in B_f$ we get

$$\lim_{N\to\infty}\frac{f(x_N)}{N} = \lim_{N\to\infty}\frac{1}{N}\left(\sum_{k=1}^{N} f(x_k) - \sum_{k=1}^{N-1} f(x_k)\right)$$
$$= \lim_{N\to\infty}\frac{1}{N}\sum_{k=1}^{N} f(x_k) - \lim_{N\to\infty}\frac{1}{N}\sum_{k=1}^{N-1} f(x_k)$$
$$= \lim_{N\to\infty}\frac{1}{N}\sum_{k=1}^{N} f(x_k) - \lim_{N-1\to\infty}\frac{N-1}{N}\left(\frac{1}{N-1}\sum_{k=1}^{N-1} f(x_k)\right)$$
$$= \lim_{N\to\infty}\frac{1}{N}\sum_{k=1}^{N} f(x_k) - \lim_{N-1\to\infty}\frac{1}{N-1}\sum_{k=1}^{N-1} f(x_k) = 0. \tag{1.4.5}$$

Remark 1.4.1 Note that for each Lebesgue integrable function f in $(0, 1)$, the following inclusion $S\cap A_f \subseteq S\cap C_f$ holds true, but the converse inclusion is not always valid. Indeed, let $(x_k)_{k\in N}$ be an arbitrary sequence of uniformly distributed numbers in (0, 1). Then the function $f : (0, 1) \to \mathbf{R}$, defined by $f(x) = \chi_{(0,1)\setminus\{x_k:k\in\mathbf{N}\}}(x)$ for $x \in (0, 1)$, is Lebesgue integrable, $(x_k)_{k\in N} \in C_f \cap S$ but $(x_k)_{k\in N} \notin A_f \cap S$ because

$$\lim_{N\to\infty} \frac{1}{N} \sum_{n=1}^{N} f(x_n) = 0 \neq 1 = \int_{(0,1)} f(x)dx. \tag{1.4.6}$$

Theorem 1.4.3 *Let $f : (0, 1) \to \mathbf{R}$ be a Lebesgue integrable function. Then the set D_f of all uniformly distributed sequences in* (0, 1) *for which the following conditions*

1. $\lim_{n\to\infty} \frac{f(x_n)}{n} = 0.$
2. $\lim_{N\to\infty} \frac{1}{N} \sum_{k=1}^{N} f(x_k)$ *exists.*
3. $\lim_{N\to\infty} \frac{1}{N} \sum_{k=1}^{N} f(x_k) = \int_{(0,1)} f(x)dx.$
4. $(x_k)_{k\in\mathbf{N}}$ *is uniformly distributed in* (0, 1)

are equivalent, has ℓ_1^∞ measure 1, *and*

$$D_f = \big(A_f \cap S\big) \cup \big((S\setminus A_f) \cap \big(S\setminus B_f\big) \cap (S\setminus C_f)\big) = \big(A_f \cap S\big) \cup (S\setminus C_f),$$

where S, A_f, B_f and C_f come from Lemma 1.2.4, Theorem 1.3.1 (when $(X, \mathbf{F}, \mu) = ((0, 1), \mathbf{B}(0, 1), \ell_1)$), and Observations 1.4.1 and 1.4.2, respectively.

Proof By Lemma 1.2.4 we know that $\ell_1^\infty(S) = 1$. By Theorem 1.3.1 we know that $\ell_1^\infty(A_f) = 1$ whenever $(X, \mathbf{F}, \mu) = ((0, 1), \mathbf{B}((0, 1)), \ell_1)$. Following Observations 1.4.1 and 1.4.2 we have $A_f \subseteq B_f \subseteq C_f$. Because $S_f = A_f \cap B_f \cap C_f \cap S = A_f \cap S$, we get

$$\ell_1^\infty(S_f) = \ell_1^\infty(A_f \cap S) = 1. \tag{1.4.7}$$

Because $S_f \subseteq D_f$ we end the proof of theorem.

Corollary 1.4.1 *Let $\mathbf{Q}$ be a set of all rational numbers of* [0, 1] *and $F \subseteq [0, 1] \cap \mathbf{Q}$ be finite. Let $f : [0, 1] \to R$ be Lebesgue integrable, and ℓ_1 almost everywhere continuous and locally bounded on $[0, 1]\setminus F$. Assume that for every $\beta \in F$ there is some neighborhood U_β of β such that f is either bounded or monotone in $[0, \beta) \cap U_\beta$ and in $(\beta, 1] \cap U_\beta$ as well. Let S, A_f, B_f and C_f come from Lemma 1.2.4, Theorem 1.3.1 (when $(X, \mathbf{F}, \mu) = ((0, 1), \mathbf{B}(0, 1), \ell_1)$), and Observations 1.4.1 and 1.4.2, respectively. We set*

$$D_f = \big(A_f \cap S\big) \cup \big((S\setminus A_f) \cap \big(S\setminus B_f\big) \cap (S\setminus C_f)\big) = \big(A_f \cap S\big) \cup (S\setminus C_f).$$

Then for $(x_k)_{k\in\mathbf{N}} \in D_f$ the following conditions are equivalent.

1. $\lim_{n\to\infty} \frac{f(x_n)}{n} = 0.$
2. $\lim_{N\to\infty} \frac{1}{N} \sum_{k=1}^{N} f(x_k)$ *exists.*
3. $\lim_{N\to\infty} \frac{1}{N} \sum_{k=1}^{N} f(x_k) = \int_{(0,1)} f(x)dx.$

Note that D_f is the maximal subset of the set S in which conditions 1 to 3 participated in the formulation of Corollary 1.4.1 are equivalent, provided that for each $(x_k)_{k\in N} \in D_f$ the sentences 1 to 3 are true or false simultaneously, and for each $(x_k)_{k\in N} \in S \backslash D_f$ the sentences 1 to 3 are not true or false simultaneously. This extends the main result of Baxa and Schoiβengeier [BS] because the class S^* of all sequences of the form $(\{n\alpha\})_{n\in N}$ is in D_f for each irrational number α, and not every element of D_f can be presented in the same form. For example,

$$(\{(n + 1/2(1 - \chi_{\{k:k\geq 2\}}(n)))\pi^{\chi_{\{k:k\geq 2\}}(n)}\})_{n\in \mathbf{N}} \in D_f \backslash S^*, \tag{1.4.8}$$

where $\{\cdot\}$ denotes the fractional part of the real number and $\chi_{\{k:k\geq 2\}}$ denotes the indicator function of the set $\{k : k \geq 2\}$.

Similarly, setting

$$E_f = A_f \cup ((0, 1)^\infty \backslash C_f)), \tag{1.4.9}$$

we get a maximal subset of $(0, 1)^\infty$ in which Conditions 1 to 3 participated in the formulation of Corollary 1.4.1 and are equivalent, provided that for each $(x_k)_{k\in N} \in E_f$ the sentences 1 to 3 are true or false simultaneously, and for each $(x_k)_{k\in N} \in (0, 1)^\infty \backslash E_f$ the sentences 1 to 3 are not true or false simultaneously.

Remark 1.4.2 Main results of Sect. 1.4 has been obtained in [P1].

References

[BS] Baxa, C., Schoiβengeier, J.: Calculation of improper integrals using $(n\alpha)$-sequences. Monatsh. Math. **135**(4), 265–277 (2002). (Dedicated to Edmund Hlawka on the occasion of his 85 th birthday)

[H1] Hardy, G., Littlewood, J.: Some problems of diophantine approximation. Acta Math. **37**(1), 193–239 (1914)

[H2] Hardy, G., Littlewood, J.: Some problems of diophantine approximation. Acta Math. **37**(1), 155–191 (1914)

[KN] Kuipers, L., Niederreiter, H.: Uniform distribution of sequences, Wiley-Interscience, New York (1974)

[N1] Nikolski, S.M.: Course of mathematical analysis (in Russian), no. 1, Moscow (1983)

[P1] Pantsulaia, G., Kiria T.: Calculation of Improper Integrals by Using Uniformly Distributed Sequences. arXiv:1507.02978

[Sh] Shiryaev, A.N.: Probability (in Russian). Izd. Nauka, Moscow (1980)

[S] Sobol, I.M.: Computation of improper integrals by means of equidistributed sequences. (Russian). Dokl. Akad. Nauk SSSR. **210**, 278–281 (1973)

[W] Weyl, H.: Úber ein Problem aus dem Gebiete der diophantischen Approximation. Marchr. Ges. Wiss. Gótingen. Math-phys. **K1**, 234–244 (1916)

Chapter 2
Infinite-Dimensional Monte Carlo Integration

2.1 Introduction

In mathematics, Monte Carlo integration is a technique for numerical integration using random numbers and a a particular Monte Carlo method numerically computes the Riemann integral. Whereas other algorithms usually evaluate the integrand at a regular grid, Monte Carlo randomly chooses points at which the integrand is evaluated. This method is particularly useful for higher-dimensional integrals. There are different methods to perform a Monte Carlo integration, such as uniform sampling, stratified sampling, and importance sampling. In this chapter we describe a certain technique for numerical calculation of infinite-dimensional integrals by using methods of the theory of uniform distribution modulo (u.d.mod) 1. Development of this theory for one-dimensional Riemann integrals was begun by Hermann Weyl's [W] celebrated theorem.

Theorem 2.1.1 ([KN], Theorem 1.1, p. 2) *The sequence $(x_n)_{n\in N}$ of real numbers is u.d. mod 1 if and only if for every real-valued continuous function f defined on the closed unit interval $\overline{I} = [0, 1]$ we have*

$$\lim_{N\to\infty} \frac{\sum_{n=1}^{N} f(\{x_n\})}{N} = \int_{\overline{I}} f(x)dx, \tag{2.1.1}$$

where $\{\cdot\}$ denotes the fractional part of the real number.

Main corollaries of this theorem were used successfully in Diophantine approximations and have applications to Monte Carlo integration (see, e.g., [H1, H2, KN]). During the last decades the methods of the theory of uniform distribution modulo one have been intensively used in various branches of mathematics as diverse as number theory, probability theory, mathematical statistics, functional analysis, topological algebra, and so on.

G. Pantsulaia, *Applications of Measure Theory to Statistics*,
DOI 10.1007/978-3-319-45578-5_2

In [P2], the concept of increasing families of finite subsets uniformly distributed in infinite-dimensional rectangles has been introduced and a certain infinite generalization of the Theorem 2.1.1 has been obtained as follows.

Theorem 2.1.2 ([P2], Theorem 3.5, p. 339) *Let* $(Y_n)_{n\in N}$ *be an of* $[0,1]^\infty$. *Then* $(Y_n)_{n\in N}$ *is uniformly distributed in the infinite-dimensional rectangle* $[0,1]^\infty$ *if and only if for every Riemann integrable function* f *on* $[0,1]^\infty$ *the following equality*

$$\lim_{n\to\infty}\frac{\sum_{y\in Y_n} f(y)}{\#(Y_n)} = \int_{[0,1]^\infty} f(x)d\lambda(x) \tag{2.1.2}$$

holds true, where λ *denotes the infinite-dimensional "Lebesgue measure" [B1].*

The purpose of the present chapter is to consider some corollaries and applications of Theorem 2.1.2. More precisely, we elaborate Monte Carlo integration for real-valued functions of infinitely many variables.

This chapter is organized as follows.

In Sect. 2.2, in terms of the "Lebesgue measure" λ [B1], we consider concepts of the uniform distribution and Riemann integrability in infinite-dimensional rectangles in R^∞ and prove infinite-dimensional versions of Lebesgue's [N] and Weyl's [W] famous results, respectively. In this section we show that if $(\alpha_n^{(k)})_{n\in N}$ is an infinite sequence of different integer numbers for every $k \in N$, then a set of all sequences $(x_k)_{k\in N}$ in R^∞ for which a sequence of increasing sets $(Y_n((x_k)_{k\in N}))_{n\in N}$ is not λ uniformly distributed on the $\prod_{k\in N}[a_k, b_k])$, where

$$Y_n((x_k)_{k\in N}) = \prod_{k=1}^{n}((\cup_{j=1}^{n}\{< \alpha_j^{(k)}x_k > (b_k - a_k)\}) + a_k) \times \prod_{k\in N\setminus\{1,\dots,n\}}\{a_k\}$$

and λ is the "Lebesgue measure" constructed by R. Baker in 1991, and is of λ measure zero, and hence shy in R^∞.

In Sect. 2.3, a Monte Carlo algorithm for estimating the value of infinite-dimensional Riemann integrals over infinite-dimensional rectangles in R^∞ is described. Furthermore, we introduce Riemann integrability for real-valued functions with respect to product measures in R^∞ and give some sufficient conditions under which a real-valued function of infinitely many real variables is Riemann integrable. We describe a Monte Carlo algorithm for computing infinite-dimensional Riemann integrals for such functions.

In Sect. 2.4, we consider some interesting applications of Monte Carlo algorithms for computing infinite-dimensional Riemann integrals described in Sect. 2.3.

2.2 Uniformly Distributed Sequences of an Increasing Family of Finite Sets in Infinite-Dimensional Rectangles

Definition 2.2.1 A bounded sequence $s_1, s_2, s_3, \ldots$ of real numbers is said to be equidistributed or uniformly distributed on an interval $[a, b]$ if for any subinterval $[c, d]$ of the $[a, b]$ we have

$$\lim_{n\to\infty} \frac{\#(\{s_1, s_2, s_3, \ldots, s_n\} \cap [c, d])}{n} = \frac{d-c}{b-a},$$

where # denotes a counting measure.

Remark 2.2.1 For $a \le c < d \le b$, let $][c, d][$ denote a subinterval of the $[a, b]$ that has one of the following forms $[c, d]$, $[c, d[$, $]c, d[$ or $]c, d]$. Then it is obvious to show that a bounded sequence $s_1, s_2, s_3, \ldots$ of real numbers is equidistributed or uniformly distributed on an interval $[a, b]$ iff, for any subinterval$][c, d][$ of the $[a, b]$, we have

$$\lim_{n\to\infty} \frac{\#(\{s_1, s_2, s_3, \ldots, s_n\}\cap][c, d][)}{n} = \frac{d-c}{b-a}.$$

Definition 2.2.2 (*Weyl* [W]) The sequence $s_1, s_2, s_3, \ldots$ is said to be equidistributed modulo 1 or uniformly distributed modulo 1 if the sequence $(s_n - [s_n])_{n\in N}$ of the fractional parts of the $(s_n)_{n\in N}$'s is equidistributed (equivalently, uniformly distributed) in the interval $[0, 1]$.

Example 2.2.1 ([KN], *Exercise 1.12, p. 16*) The sequence of all multiples of an irrational α

$$0, \alpha, 2\alpha, 3\alpha, \ldots$$

is uniformly distributed modulo 1.

Example 2.2.2 ([KN], Exercise 1.13, p. 16) The sequence

$$\frac{0}{1}, \frac{0}{2}, \frac{1}{2}, \frac{0}{3}, \frac{1}{3}, \frac{2}{3}, \ldots, \frac{0}{k}, \ldots, \frac{k-1}{k}, \ldots$$

is uniformly distributed modulo 1.

Example 2.2.3 The sequence of all multiples of an irrational α by successive prime numbers

$$2\alpha, 3\alpha, 5\alpha, 7\alpha, 11\alpha, \ldots$$

is equidistributed modulo 1. This is a famous theorem of analytic number theory, proved by I. M. Vinogradov in 1935 (see [V]).

Agreement In the sequel, unlike N. Bourbaki's well-known notion, under N we understand a set $\{1, 2, \dots\}$.

Remark 2.2.2 If $(s_k)_{k\in N}$ is uniformly distributed modulo 1, then $((s_k - [s_k])(b - a) + a)_{k\in N}$ is uniformly distributed in an interval $[a, b)$.

The following assertion contains an interesting application of uniformly distributed sequences for a calculation of Riemann integrals.

Lemma 2.2.1 (Weyl [W]) *The following two conditions are equivalent.*
(i) $(a_n)_{n\in N}$ *is equidistributed modulo* 1.
(ii) for every Riemann integrable function f *on* $[0, 1]$

$$\lim_{n\to\infty} \frac{1}{n}\sum_{j=1}^{n} f(a_j) = \int_{[0,1]} f(x)dx.$$

Remark 2.2.3 Let $s_1, s_2, s_3, \dots$ be uniformly distributed in an interval $[a, b]$. Setting $Y_n = \{s_1, s_2, s_3, \dots, s_n\}$ for $n \in N$, the $(Y_n)_{n\in N}$ will be an increasing sequence of finite subsets of the $[a, b]$ such that, for any subinterval $[c, d]$ of the $[a, b]$, the following equality

$$\lim_{n\to\infty} \frac{\#(Y_n \cap [c, d])}{\#(Y_n)} = \frac{d - c}{b - a}$$

will be valid.

Remark 2.2.3 gives rise to the following definition.

Definition 2.2.3 An increasing sequence $(Y_n)_{n\in N}$ of finite subsets of the $[a, b]$ is said to be equidistributed or uniformly distributed in an interval $[a, b]$ if, for any subinterval $[c, d]$ of the $[a, b]$, we have

$$\lim_{n\to\infty} \frac{\#(Y_n \cap [c, d])}{\#(Y_n)} = \frac{d - c}{b - a}.$$

Definition 2.2.4 Let $\prod_{k\in N}[a_k, b_k] \in \mathscr{R}$. A set U is called an elementary rectangle in the $\prod_{k\in N}[a_k, b_k]$ if it admits the representation:

$$U = \prod_{k=1}^{m}]c_k, d_k[\times \prod_{k\in N\setminus\{1,\dots,m\}} [a_k, b_k],$$

where $a_k \le c_k < d_k \le b_k$ for $1 \le k \le m$.

It is obvious that

$$\lambda(U)=\prod_{k=1}^{m}(d_k-c_k)\times\prod_{k=m+1}^{\infty}(b_k-a_k),$$

for the elementary rectangle U.

Definition 2.2.5 An increasing sequence $(Y_n)_{n\in N}$ of finite subsets of the infinite-dimensional rectangle $\prod_{k\in N}[a_k,b_k]\in\mathscr{R}$ is said to be uniformly distributed in the $\prod_{k\in N}[a_k,b_k]$ if for every elementary rectangle U in the $\prod_{k\in N}[a_k,b_k[$ we have

$$\lim_{n\to\infty}\frac{\#(Y_n\cap U)}{\#(Y_n)}=\frac{\lambda(U)}{\lambda(\prod_{k\in N}[a_k,b_k[)}.$$

Theorem 2.2.1 ([P2], Theorem 3.1, p. 4) *Let $\prod_{k\in N}[a_k,b_k]\in\mathscr{R}$. Let $(x_n^{(k)})_{n\in N}$ be uniformly distributed in the interval $[a_k,b_k]$ for $k\in N$. We set*

$$Y_n=\prod_{k=1}^{n}(\cup_{j=1}^{n}x_j^{(k)})\times\prod_{k\in N\setminus\{1,\dots,n\}}\{x_1^{(k)}\}.$$

Then $(Y_n)_{n\in N}$ is uniformly distributed in the rectangle $\prod_{k\in N}[a_k,b_k]$.

Proof Let U be an elementary rectangle in the $\prod_{k\in N}[a_k,b_k]$.

Because $(x_n^{(k)})_{n\in N}$ is uniformly distributed in the interval $[a_k,b_k]$ for $k\in N$, we get

$$\lim_{n\to\infty}\frac{\#(\{x_1^{(k)},x_2^{(k)},\dots,x_n^{(k)}\}\cap][c_k,d_k][)}{n}=\frac{d_k-c_k}{b_k-a_k}.$$

Therefore we have

$$\lim_{n\to\infty}\frac{\#(Y_n\cap U)}{\#(Y_n)}=\lim_{n\to\infty}\prod_{k=1}^{m}\frac{\#(\{x_1^{(k)},x_2^{(k)},\dots,x_n^{(k)}\}\cap][c_k,d_k][)}{n}$$
$$=\prod_{k=1}^{m}\lim_{n\to\infty}\frac{\#(\{x_1^{(k)},x_2^{(k)},\dots,x_n^{(k)}\}\cap][c_k,d_k][)}{n}$$
$$=\prod_{k=1}^{m}\frac{d_k-c_k}{b_k-a_k}=\frac{\lambda(U)}{\lambda(\prod_{k\in N}[a_k,b_k])}. \tag{2.2.1}$$

The theorem is proved.

Remark 2.2.4 In the context of Theorem 2.2.1, it is natural to ask whether there exists an increasing sequence of finite subsets $(Y_n)_{n\in N}$ such that

$$\lim_{n\to\infty}\frac{\#(Y_n\cap U)}{\#(Y_n)}=\frac{\lambda(U)}{\lambda(\prod_{k\in N}[a_k,b_k])}$$

for every infinite-dimensional rectangle $U=\prod_{k\in N}X_k\subset\prod_{k\in N}[a_k,b_k]$, where, for each $k\in N$, X_k is a finite sum of pairwise disjoint intervals in $[a_k,b_k]$.

Let us show that the answer to this question is no.

Indeed, assume the contrary and let $(Y_n)_{n\in N}$ be such an increasing sequence of finite subsets in $\prod_{k\in N}[a_k,b_k[$. Then we have

$$\cup_{n\in N}Y_n=\{(x_i^{(k)})_{i\in N}:k\in N\}.$$

For $k\in N$, we set $X_k=[a_k,b_k]\setminus x_k^{(k)}$. Then, it is clear that

$$\lambda\Big(\prod_{k\in N}X_k\Big)=\lambda\Big(\prod_{k\in N}[a_k,b_k]\Big)$$

and

$$\frac{\#(Y_n\cap\prod_{k\in N}X_k)}{\#(Y_n)}=0$$

for $k\in N$, which follows

$$\lim_{n\to\infty}\frac{\#(Y_n\cap\prod_{k\in N}X_k)}{\#(Y_n)}=0<1=\frac{\lambda(\prod_{k\in N}X_k)}{\lambda(\prod_{k\in N}[a_k,b_k])}.$$

Definition 2.2.6 Let $\prod_{k\in N}[a_k,b_k]\in\mathscr{R}$. A family of pairwise disjoint elementary rectangles $\tau=(U_k)_{1\le k\le n}$ of the $\prod_{k\in N}[a_k,b_k]$ is called the Riemann partition of the $\prod_{k\in N}[a_k,b_k]$ if $\cup_{1\le k\le n}U_k=\prod_{k\in N}[a_k,b_k]$.

Definition 2.2.7 Let $\tau=(U_k)_{1\le k\le n}$ be the Riemann partition of the $\prod_{k\in N}[a_k,b_k]$. Let $\ell(Pr_i(U_k))$ be a length of the ith projection $Pr_i(U_k)$ of the U_k for $i\in N$. We set

$$d(U_k)=\sum_{i\in N}\frac{\ell(Pr_i(U_k))}{2^i(1+\ell(Pr_i(U_k))}.$$

It is obvious that $d(U_k)$ is a diameter of the elementary rectangle U_k for $k\in N$ with respect to the Tikhonov metric ρ defined as

$$\rho((x_k)_{k\in N},(y_k)_{k\in N})=\sum_{k\in N}\frac{|x_k-y_k|}{2^k(1+|x_k-y_k|)}$$

for $(x_k)_{k\in N},(y_k)_{k\in N}\in\mathbf{R}^\infty$.

A number $d(\tau)$, defined by

$$d(\tau) = \max\{d(U_k) : 1 \le k \le n\}$$

is called the mesh or norm of the Riemann partition τ.

Definition 2.2.8 Let $\tau_1 = (U_i^{(1)})_{1 \le i \le n}$ and $\tau_2 = (U_j^{(2)})_{1 \le j \le m}$ be Riemann partitions of the $\prod_{k \in N}[a_k, b_k]$. We say that $\tau_2 \le \tau_1$ iff

$$(\forall j)((1 \le j \le m) \rightarrow (\exists i_0)(1 \le i_0 \le n \;\&\; U_j^{(2)} \subseteq U_{i_0}^{(1)})).$$

Definition 2.2.9 Let f be a real-valued bounded function defined on the $\prod_{i \in N}[a_i, b_i]$. Let $\tau = (U_k)_{1 \le k \le n}$ be the Riemann partition of the $\prod_{k \in N}[a_k, b_k]$ and $(t_k)_{1 \le k \le n}$ be a sample such that, for each k, $t_k \in U_k$. Then

(i) A sum $\sum_{k=1}^{n} f(t_k)\lambda(U_k)$ is called the Riemann sum of the f with respect to Riemann partition $\tau = (U_k)_{1 \le k \le n}$ together with sample $(t_k)_{1 \le k \le n}$.

(ii) A sum $S_\tau = \sum_{k=1}^{n} M_k \lambda(U_k)$ is called the upper Darboux sum with respect to Riemann partition τ, where $M_k = \sup_{x \in U_k} f(x)(1 \le k \le n)$.

(iii) A sum $s_\tau = \sum_{k=1}^{n} m_k \lambda(U_k)$ is called the lower Darboux sum with respect to Riemann partition τ, where $m_k = \inf_{x \in U_k} f(x)(1 \le k \le n)$.

Definition 2.2.10 Let f be a real-valued bounded function defined on $\prod_{i \in N}[a_i, b_i[$. We say that the f is Riemann integrable on $\prod_{i \in N}[a_i, b_i]$ if there exists a real number s such that for every positive real number ε there exists a real number $\delta > 0$ such that, for every Riemann partition $(U_k)_{1 \le k \le n}$ of the $\prod_{k \in N}[a_k, b_k]$ with $d(\tau) < \delta$ and for every sample $(t_k)_{1 \le k \le n}$, we have

$$\Big|\sum_{k=1}^{n} f(t_k)\lambda(U_k) - s\Big| < \varepsilon.$$

The number s is called the Riemann integral and is denoted by

$$(R)\int_{\prod_{k \in N}[a_k, b_k]} f(x) d\lambda(x).$$

Definition 2.2.11 A function f is called a step function on $\prod_{k \in N}[a_k, b_k]$ if it can be written as

$$f(x) = \sum_{k=1}^{n} c_k \mathscr{X}_{U_k}(x),$$

where $\tau = (U_k)_{1 \le k \le n}$ is any Riemann partition of the $\prod_{k \in N}[a_k, b_k]$, $c_k \in R$ for $1 \le k \le n$ and $\mathscr{X}_A$ is the indicator function of the A.

Theorem 2.2.2 *Let f be a continuous function on $\prod_{k\in N}[a_k, b_k]$ with respect to the Tikhonov metric ρ. Then f is Riemann integrable on $\prod_{k\in N}[a_k, b_k]$.*

Proof It is obvious that, for every Riemann partition $\tau = (U_k)_{1\leq k\leq n}$ of the $\prod_{k\in N}[a_k, b_k]$ and for every sample $(t_k)_{1\leq k\leq n}$ with $t_k \in U_k (1 \leq k \leq n)$, we have

$$s_\tau \leq \sum_{k=1}^{n} f(t_k)\lambda(U_k) \leq S_\tau.$$

Note that if τ_1 and τ_2 are two Riemann partitions of the $\prod_{k\in N}[a_k, b_k]$ such that $\tau_2 \leq \tau_1$, then

$$s_{\tau_1} \leq s_{\tau_2} \leq \sum_{k=1}^{n} f(t_k)\lambda(U_k) \leq S_{\tau_2} \leq S_{\tau_1}.$$

Let us show the validity of the condition

$$(\forall \varepsilon)(\varepsilon > 0 \rightarrow (\exists r)(\forall \tau)(d(\tau) < r \rightarrow S_\tau - s_\tau < \varepsilon)),$$

which yields that $\inf_\tau S_\tau = \sup_\tau s_\tau$.

Following Tikhonov's theorem, the $\prod_{k\in N}[a_k, b_k]$ is a compact set in the Polish group $\mathbf{R}^\infty$ equipped with the Tikhonov metric ρ.

Following Cantor's well-known result, the function f is uniformly continuous on the $\prod_{k\in N}[a_k, b_k]$. Thus for $\varepsilon > 0$, there exists $r > 0$ such that

$$(\forall x, y)\Big(x, y \in \prod_{k\in N}[a_k, b_k]\&\rho(x, y) < r \rightarrow |f(x) - f(y)| \leq \frac{\varepsilon}{\lambda(\prod_{k\in N}[a_k, b_k])}\Big).$$

Thus, for every Riemann partition $\tau = (U_k)_{1\leq k\leq n}$ with $d(\tau) < r$ we get

$$S_\tau - s_\tau \leq \frac{\varepsilon}{\lambda(\prod_{k\in N}[a_k, b_k[)} \times \sum_{1\leq k\leq n} \lambda(U_k) = \varepsilon.$$

Thus, $\inf_\tau S_\tau = \sup_\tau s_\tau$.

Finally, setting $\delta = r$ and $s = \inf_\tau S_\tau$, we deduce that for every Riemann partition $(U_k)_{1\leq k\leq n}$ of the $\prod_{k\in N}[a_k, b_k]$ with $d(\tau) < \delta$ and for every sample $(t_k)_{1\leq k\leq n}$ with $t_k \in U_k (1 \leq k \leq n)$, we have

$$\left|\sum_{k=1}^{n} f(t_k)\lambda(U_k) - s\right| \leq S_\tau - s_\tau \leq \varepsilon.$$

This ends the proof of Theorem 2.2.2.

We have the following infinite-dimensional version of the Lebesgue theorem (see [N], Lebesgue Theorem, p. 359).

Theorem 2.2.3 *Let f be a bounded real-valued function on $\prod_{k\in N}[a_k, b_k] \in \mathcal{R}$. Then f is Riemann integrable on $\prod_{k\in N}[a_k, b_k]$ if and only if f is λ almost continuous on $\prod_{k\in N}[a_k, b_k]$.*

Proof (*Necessity*) Let f be a Riemann integrable function on $\prod_{k\in N}[a_k, b_k] \in \mathcal{R}$.

Then, for every $\varepsilon > 0$ and $\mu > 0$, there exists a Riemann partition $\tau = (U_k)_{1\le k\le n}$ such that

$$\varepsilon \times \mu \ge S_\tau - s_\tau \ge \sum_{1\le k\le n}(M_k - m_k)\lambda(U_k)$$
$$\ge \sum_{k\in I_1}(M_k - m_k)\lambda(U_k) \ge \mu \sum_{k\in I_1}\lambda(U_k), \tag{2.2.1}$$

where $I_1 = \{k : 1 \le k \le n \;\&\; U_k$ contains at least one inner point p belonging to the set $E_\mu\}$, where

$$E_\mu = \left\{x : x \in \prod_{k\in N}[a_k, b_k] \;\&\; \omega(f, x) \ge \mu\right\}$$

and

$$\omega(f, x) = \lim_{\delta\to 0} \sup_{x',x''\in V(x,\delta)\cap\prod_{k\in N}[a_k,b_k]} |f(x') - f(x'')|.$$

Here, for $x \in \mathbf{R}^\infty$ and $\delta > 0$, $V(x, \delta)$ is denoted by

$$V(x, \delta) = \left\{y : y \in \prod_{k\in N}[a_k, b_k] \;\&\rho(x, y) \le \delta\right\}.$$

Because, for $k \in I_1$, p is an inner point of the U_k, there exists $V(p, \delta(k, p))$ such that $V(p, \delta(k, p)) \subseteq U_k$.

Inasmuch as $\omega(f, p) \ge \mu$, we have

$$M_k - m_k \ge M_p - m_p \ge \omega(f, p) \ge \mu,$$

where

$$M_p = \sup_{x\in V(p,\delta(k,p))} f(x), \; m_\delta = \inf_{x\in V(p,\delta(k,p))} f(x).$$

From (2.2.1), we get

$$\varepsilon \geq \sum_{k \in I_1} \lambda(U_k).$$

Other points of the E_μ, which are not inner points of elements of the partition τ, may be placed on the boundary of elements of the τ, whose λ measure is zero.

Thus, for $\mu > 0$, we have

$$\lambda(E_\mu) \leq \sum_{k \in I_1} \lambda(U_k) + \lambda(\cup_{1 \leq k \leq n} \partial(U_k)) \leq \frac{\varepsilon}{\mu},$$

which yields that $\lambda(E_\mu) = 0$. Because a set E of all points of discontinuity of the f admits the representation $E = \cup_{k=1}^{\infty} E_{\frac{1}{k}}$, we deduce that $\lambda(E) = 0$.

This ends Necessity.

Proof of the sufficiency. Let, for $K \in \mathbf{R}^+$, us have $|f(x)| \leq K$ whenever $x \in \prod_{k \in N}[a_k, b_k]$.

Suppose that f is λ almost continuous on $\prod_{k \in N}[a_k, b_k]$.

For $\varepsilon > 0$, let μ be such a positive number that

$$4\mu\lambda\Big(\prod_{k \in N}[a_k, b_k]\Big) < \varepsilon.$$

Because, for a set E of all points of discontinuity of the f on $\prod_{k \in N}[a_k, b_k]$ we have $\lambda(E) = 0$, we easily claim that $\lambda(E_\mu) = 0$. Because E_μ is closed in $\prod_{k \in N}[a_k, b_k]$, we claim that E_μ is compact. Hence, for $\varepsilon > 0$, there exists a finite family of open elementary rectangles in $\prod_{k \in N}[a_k, b_k]$ whose union covers E_μ such that

$$\lambda(\cup_{1 \leq k \leq n} U_k) < \frac{\varepsilon}{4K}.$$

Finally, we have

$$\prod_{k \in N}[a_k, b_k] = \cup_{1 \leq k \leq n} U_k \cup F,$$

where F is a compact subset in $\prod_{k \in N}[a_k, b_k]$.

It is obvious that, for every point $x \in F$, we have $\omega(f, x) < \mu$. Because F is compact, we can choose $\delta > 0$ such that for every $x, x^{'} \in F$ the condition $\rho(x, x^{'}) < \delta$ yields $\left|f(x) - f(x^{'})\right| < 2\lambda$.

F is a finite union of elementary rectangles in $\prod_{k \in N}[a_k, b_k]$ (this follows from the fact that the class of all elementary rectangles in $\prod_{k \in N}[a_k, b_k]$ is a ring), therefore there exists a partition $\tau_1 = (F_i)_{2 \leq i \leq m}$ of the F such that, for i with $2 \leq i \leq m$, F_i is an elementary rectangle in $\prod_{k \in N}[a_k, b_k]$ with $d(F_i) < \delta$. Then $\tau = \{\cup_{1 \leq k \leq n} U_k, F_2, \ldots, F_m\}$ will be a Riemann partition of the $\prod_{k \in N}[a_k, b_k]$ such that

$$S_\tau - s_\tau = (M_1 - m_1)\lambda(\cup_{1 \le k \le n} U_k) + \sum_{1 \le i \le m} (M_i - m_i)\lambda(F_k)$$
$$\le \frac{\varepsilon}{2} + 2\mu\lambda(\prod_{k \in N}[a_k, b_k])) \le \frac{\varepsilon}{2} + \frac{\varepsilon}{2} = \varepsilon.$$

The theorem is proved.

Remark 2.2.5 Note that Theorem 2.2.2 can be considered as a simple consequence of Theorem 2.2.3. Now, by using Theorem 2.2.3, one can extend the concept of the Riemann integrability theory for functions defined in the topological vector space $\mathbf{R}^\infty$ of all real-valued sequences equipped with Tikhonov topology.

In the sequel we need some important notions and well-known results from general topology and measure theory.

Definition 2.2.12 A topological Hausdorff space X is called normal if given any disjoint closed sets E and F, there are neighborhoods U of E and V of F that are also disjoint.

Lemma 2.2.2 (Urysohn [Ur]) *A topological space X is normal if and only if any two disjoint closed sets can be separated by a function. That is, given disjoint closed sets E and F, there is a continuous function f from X to $[0, 1]$ such that the preimages of 0 and 1 under f are E and F, respectively.*

Remark 2.2.6 Because all compact Hausdorff spaces are normal, we deduce that $\prod_{k \in N}[a_k, b_k]$ equipped with Tikhonov topology, is normal. By Urysohn's lemma we deduce that any two disjoint closed sets in $\prod_{k \in N}[a_k, b_k]$ can be separated by a function.

Definition 2.2.13 A Borel measure μ, defined on a Hausdorff topological space X, is called a Radon if

$$(\forall Y)(Y \in \mathscr{B}(X) \,\&\, 0 \le \mu(Y) < +\infty \rightarrow \mu(Y) = \sup_{\substack{K \subseteq Y \\ \text{K is compact in X}}} \mu(K)).$$

Lemma 2.2.3 (Ulam [Ul]) *Every probability Borel measure defined on Polish metric space is a Radon.*

In the sequel we denote by $\mathscr{C}(\prod_{k \in N}[a_k, b_k])$ a class of all continuous (with respect to Tikhonov topology) real-valued functions on $\prod_{k \in N}[a_k, b_k]$.

Theorem 2.2.4 *For $\prod_{i\in N}[a_i,b_i]\in\mathcal{R}$, let $(Y_n)_{n\in N}$ be an increasing family of its finite subsets. Then $(Y_n)_{n\in N}$ is uniformly distributed in the $\prod_{k\in N}[a_k,b_k]$ if and only if for every $f\in\mathcal{C}(\prod_{k\in N}[a_k,b_k])$ the equality*

$$\lim_{n\to\infty}\frac{\sum_{y\in Y_n}f(y)}{\#(Y_n)}=\frac{(R)\int_{\prod_{k\in N}[a_k,b_k]}f(x)d\lambda(x)}{\lambda\left(\prod_{i\in N}[a_i,b_i]\right)}$$

holds.

Proof **Necessity**. Let $(Y_n)_{n\in N}$ be a uniformly distributed in the $\prod_{k\in N}[a_k,b_k]$ and let $f(x)=\sum_{k=1}^m c_k\mathcal{X}_{U_k}(x)$ be a step function. Then we have

$$\lim_{n\to\infty}\frac{\sum_{y\in Y_n}f(y)}{\#(Y_n)}=\lim_{n\to\infty}\frac{\sum_{y\in Y_n}\sum_{k=1}^m c_k\mathcal{X}_{U_k}(y)}{\#(Y_n)}$$

$$=\lim_{n\to\infty}\frac{\sum_{k=1}^m c_k\#(U_k\cap Y_n)}{\#(Y_n)}=\sum_{k=1}^m c_k\lim_{n\to\infty}\frac{\#(U_k\cap Y_n)}{\#(Y_n)}$$

$$=\sum_{k=1}^m c_k\frac{\lambda(U_k)}{\lambda(\prod_{i\in N}[a_i,b_i])}=\frac{(R)\int_{\prod_{k\in N}[a_k,b_k]}f(x)d\lambda(x)}{\lambda(\prod_{i\in N}[a_i,b_i])}.$$

Now, let $f\in\mathcal{C}(\prod_{k\in N}[a_k,b_k])$. By Theorem 2.2.3 we deduce that f is Riemann integrable. From the definition of the Riemann integral we deduce that, for every positive ε, there exist two step functions f_1 and f_2 on $\prod_{i\in N}[a_i,b_i]$ such that

$$f_1(x)\le f(x)\le f_2(x)$$

and

$$(R)\int_{\prod_{i\in N}[a_i,b_i]}(f_1(x)-f_2(x))d\lambda(x)<\varepsilon.$$

Then we have

$$(R)\int_{\prod_{i\in N}[a_i,b_i]}f(x)d\lambda(x)-\varepsilon\le(R)\int_{\prod_{i\in N}[a_i,b_i]}f_1(x)d\lambda(x)$$

$$=\lambda\Big(\prod_{i\in N}[a_i,b_i]\Big)\times\lim_{n\to\infty}\frac{\sum_{y\in Y_n}f_1(y)}{\#(Y_n)}\le\lambda\Big(\prod_{i\in N}[a_i,b_i]\Big)\times\underline{\lim}_{n\to\infty}\frac{\sum_{y\in Y_n}f(y)}{\#(Y_n)}$$

$$\le\lambda\Big(\prod_{i\in N}[a_i,b_i]\Big)\times\overline{\lim}_{n\to\infty}\frac{\sum_{y\in Y_n}f(y)}{\#(Y_n)}\le\lim_{n\to\infty}\frac{\sum_{y\in Y_n}f_2(y)}{\#(Y_n)}$$

$$\leq \lambda\Big(\prod_{i\in N}[a_i,b_i]\Big) \times (R)\int_{\prod_{i\in N}[a_i,b_i]} f_2(x)d\lambda(x) \leq (R)\int_{\prod_{i\in N}[a_i,b_i]} f(x)d\lambda(x)+\varepsilon.$$

The latter relation yields an existence of the limit $\lim_{n\to\infty}\frac{\sum_{y\in Y_n} f(y)}{\#(Y_n)}$ such that

$$\lim_{n\to\infty}\frac{\sum_{y\in Y_n} f(y)}{\#(Y_n)} = \frac{(R)\int_{\prod_{k\in N}[a_k,b_k]} f(x)d\lambda(x)}{\lambda(\prod_{i\in N}[a_i,b_i])}.$$

This ends the proof of Necessity.

Sufficiency. Assume that $(Y_n)_{n\in N}$ is an increasing sequence of subsets of the $\prod_{k\in N}[a_k,b_k]$ such that for every $f\in\mathscr{C}(\prod_{k\in N}[a_k,b_k])$ the equality

$$\lim_{n\to\infty}\frac{\sum_{y\in Y_n} f(y)}{\#(Y_n)} = \frac{(R)\int_{\prod_{k\in N}[a_k,b_k[} f(x)d\lambda(x)}{\lambda(\prod_{i\in N}[a_i,b_i])}$$

holds.

Let U be any elementary rectangle in $\prod_{i\in N}[a_i,b_i]$.

For $\varepsilon>0$, by Ulam's lemma we can choose such a compact set

$$F\subset\prod_{k\in N}[a_k,b_k]\setminus[U]_T,$$

that $\lambda((\prod_{k\in N}[a_k,b_k]\setminus[U]_T)\setminus F)<\frac{\varepsilon}{2}$, where $[U]_T$ denotes completion of the set U by Tikhonov topology in $\prod_{k\in N}[a_k,b_k]$. Then, by Urysohn's lemma we deduce that there is a continuous function g_2 from $\prod_{k\in N}[a_k,b_k]$ to $[0,1]$ such that the preimages of 0 and 1 under g_2 are F and $[U]_T$, respectively. Then, for $x\in\prod_{k\in N}[a_k,b_k]$, we have

$$\mathscr{X}_U(x)\leq g_2(x)$$

and

$$(R)\int_{\prod_{k\in N}[a_k,b_k]}(g_2(x)-\mathscr{X}_U(x))d\lambda(x)\leq\frac{\varepsilon}{2},$$

where $\mathscr{X}_U$ is an indicator of the U defined on the $\prod_{k\in N}[a_k,b_k]$.

Now let us consider a set $[\prod_{k\in N}[a_k,b_k]\setminus U]_T$. Using Ulam's lemma, we can choose such a compact set

$$F_1\subset\prod_{k\in N}[a_k,b_k]\setminus\Big[\prod_{k\in N}[a_k,b_k]\setminus U\Big]_T$$

that

$$\lambda\Big(\Big(\prod_{k\in N}[a_k,b_k]\setminus\Big[\prod_{k\in N}[a_k,b_k]\setminus U\Big]_T\Big)\setminus F_1\Big)<\frac{\varepsilon}{2}.$$

Then, by Urysohn's lemma we deduce that there is a continuous function g_1 from $\prod_{k\in N}[a_k,b_k]$ to $[0,1]$ such that the preimages of 0 and 1 under g_1 are $[\prod_{k\in N}[a_k,b_k]\setminus U]_T$ and F_1, respectively. Then, for $x\in\prod_{k\in N}[a_k,b_k]$, we have

$$g_1(x)\le\mathscr{X}_U(x)$$

and

$$(R)\int_{\prod_{k\in N}[a_k,b_k]}(\mathscr{X}_U(x)-g_1(x))d\lambda(x)\le\frac{\varepsilon}{2}.$$

Now, we deduce that for every elementary rectangle U in $\prod_{i\in N}[a_i,b_i]$ there exist two continuous functions g_1 and g_2 on the $\prod_{i\in N}[a_i,b_i]$ such that

$$g_1(x)\le\mathscr{X}_U(x)\le g_2(x)$$

and

$$(R)\int_{\prod_{i\in N}[a_i,b_i]}(g_2(x)-g_1(x))d\lambda(x)\le\varepsilon.$$

Then we have

$$\lambda(U)-\varepsilon\le(R)\int_{\prod_{i\in N}[a_i,b_i]}g_2(x)d\lambda(x)-\varepsilon\le(R)\int_{\prod_{i\in N}[a_i,b_i]}g_1(x)d\lambda(x)$$

$$=\lambda\Big(\prod_{i\in N}[a_i,b_i]\Big)\times\lim_{n\to\infty}\frac{\sum_{y\in Y_n}g_1(y)}{\#(Y_n)}\le\lambda\Big(\prod_{i\in N}[a_i,b_i]\Big)\times\underline{\lim}_{n\to\infty}\frac{\#(Y_n\cap U)}{\#(Y_n)}$$

$$\le\lambda\Big(\prod_{i\in N}[a_i,b_i]\Big)\times\overline{\lim}_{n\to\infty}\frac{\#(Y_n\cap U)}{\#(Y_n)}\le\lambda\Big(\prod_{i\in N}[a_i,b_i]\Big)\times\lim_{n\to\infty}\frac{\sum_{y\in Y_n}g_2(y)}{\#(Y_n)}$$

$$=(R)\int_{\prod_{i\in N}[a_i,b_i]}g_2(x)d\lambda(x)\le(R)\int_{\prod_{i\in N}[a_i,b_i]}g_1(x)d\lambda(x)+\varepsilon\le\lambda(U)+\varepsilon.$$

Because ε was taken as arbitrary, we deduce that

$$\lambda\Big(\prod_{i\in N}[a_i,b_i]\Big)\times\lim_{n\to\infty}\frac{\#(Y_n\cap U)}{\#(Y_n)}=\lambda(U).$$

This ends the proof of Theorem 2.2.4.

Now by the scheme used in the proof of Theorem 2.2.4, one can get the validity of an infinite-dimensional analogue of Lemma 2.2.1. In particular, the following assertion is valid.

Theorem 2.2.5 *For $\prod_{i\in N}[a_i, b_i] \in \mathscr{R}$, let $(Y_n)_{n\in N}$ be an increasing family of its finite subsets. Then $(Y_n)_{n\in N}$ is uniformly distributed in the $\prod_{k\in N}[a_k, b_k]$ if and only if for every Riemann integrable function f on $\prod_{k\in N}[a_k, b_k]$ the equality*

$$\lim_{n\to\infty}\frac{\sum_{y\in Y_n} f(y)}{\#(Y_n)} = \frac{(R)\int_{\prod_{k\in N}[a_k,b_k]} f(x)d\lambda(x)}{\lambda(\prod_{i\in N}[a_i, b_i])}$$

holds.

Definition 2.2.14 (*Weyl* [W]) A sequence $s_1, s_2, s_3, \ldots$ is said to be equidistributed modulo 1 or uniformly distributed modulo 1 if the sequence $(< s_n >)_{n\in N}$ of the fractional parts of $(s_n)_{n\in N}$'s is equidistributed (equivalently, uniformly distributed) on the interval $[0, 1]$.

Lemma 2.2.4 *Let $\prod_{k\in N}[a_k, b_k] \in \mathscr{R}$. Let $(x_n^{(k)})_{n,k\in N}$ be a double sequence of elements of $\prod_{k\in N}[a_k, b_k]$. We set*

$$Y_n = \prod_{k=1}^{n}(\cup_{j=1}^{n}\{x_j^{(k)}\}) \times \prod_{k\in N\setminus\{1,\ldots,n\}}\{x_1^{(k)}\}.$$

Then $(Y_n)_{n\in N}$ is uniformly distributed in the rectangle $\prod_{k\in N}[a_k, b_k]$ if and only if $(x_n^{(k)})_{n\in N}$ is uniformly distributed on the interval $[a_k, b_k]$ for $k \in N$.

Proof (*Sufficiency*) Because $(Y_n)_{n\in N}$ is uniformly distributed in the rectangle $\prod_{k\in N}[a_k, b_k]$, for an elementary rectangle $U = \prod_{i=1}^{k-1}[a_i, b_i]\times][c, d][\times \prod_{i=k+1}^{+\infty}[a_i, b_i]$ with $][c, d][\subseteq [a_k, b_k]$, we have

$$\frac{d-c}{b_k-a_k} = \frac{\lambda(U)}{\lambda(\prod_{i\in N}[a_i, b_i[)} = \lim_{n\to\infty}\frac{\#(Y_n\cap U)}{\#(Y_n)} = \lim_{n\to\infty}\frac{\#(\{x_1^{(k)},\ldots,x_n^{(k)}\}\cap U)}{n}.$$

The latter relation means that $(x_n^{(k)})_{n\in N}$ is uniformly distributed on the interval $[a_k, b_k]$ for $k \in N$.

Necessity. See Theorem 2.2.1.

Lemma 2.2.5 ([KN], Theorem 4.1, p. 42) *Let $(\alpha_n)_{n\in N}$ be a sequence of different integer numbers. Then a sequence of real numbers $(\alpha_n x)_{n\in N}$ is u.d. mod 1 for l_1 almost points of R.*

Definition 2.2.15 Following Brian R. Hunt, Tim Sauer, and James A. Yorke (cf. [HSY]), a set X is called shy if it is a subset of a Borel set X' for which $\mu(X'+v)=0$ for every $v \in B$ and some Borel probability measure μ such that $\mu(K)=\mu(B)$ for some compact K.

Definition 2.2.16 A Borel measure μ in V is called a generator (of shy sets) in V, if

$$(\forall X)(\overline{\mu}(X)=0 \rightarrow X \in S(V)),$$

where $\overline{\mu}$ denotes a usual completion of the Borel measure μ.

Lemma 2.2.6 ([P4], Example 2.1, p. 242) *"Lebesgue measure" [B1] is a generator of shy sets.*

Theorem 2.2.6 *Let $(\alpha_n^{(k)})_{n \in N}$ be an infinite sequence of different integer numbers for every $k \in N$. Then a set of all sequences $(x_k)_{k \in N}$ in R^∞ for which a sequence of increasing sets $(Y_n((x_k)_{k \in N}))_{n \in N}$ is not λ u.d. on the $\prod_{k \in N}[a_k, b_k]$, where*

$$Y_n((x_k)_{k \in N}) = \prod_{k=1}^{n}(\cup_{j=1}^{n}\{< \alpha_j^{(k)} x_k > (b_k - a_k)\}) \times \prod_{k \in N \setminus \{1,\dots,n\}} \{a_k\},$$

is of λ measure zero, where $< \cdot >$ denotes the fractal part of the number.

Proof For $k \in N$, we denote D_k by

$D_k = \{x_k : x_k \in R \;\&\; (< \alpha_j^{(k)} x_k > (b_k - a_k))_{j \in N}$ is l_1 u.d. on $[a_k, b_k]\}$.

By Lemma 2.2.4 we have that $l_1(R \setminus D_k) = 0$ for $k \in N$.

We set $D = \prod_{k \in N} D_k$. It is clear that $\lambda(R^\infty \setminus D) = 0$. For $(x_k)_{k \in N} \in D$, we have that $(< \alpha_j^{(k)} x_k > (b_k - a_k))_{j \in N}$ is l_1 u.d. on $[a_k, b_k]$ for every $k \in N$. By Lemma 2.2.4 we claim that $(Y_n((x_k)_{k \in N}))_{n \in N}$ is λ u.d. on the $\prod_{k \in N}[a_k, b_k]$ for $(x_k)_{k \in N} \in D$. Inasmuch as $\lambda(R^\infty \setminus D) = 0$, Theorem 2.2.6 is proved.

Remark 2.2.7 Following Lemma 2.2.6, λ is the generator of shy sets. The latter relation means that every set of λ measure zero is shy in R^∞. Following Theorem 2.2.6, we claim that a set of all sequences $(x_j)_{j \in N}$ in R^∞ for which a sequence of increasing sets $(Y_n((x_j)_{j \in N}))_{n \in N}$ is λ u.d. on the $\prod_{k \in N}[a_k, b_k])$ is the prevalent set.

2.3 Monte Carlo Algorithm for Estimating the Value of Infinite-Dimensional Riemann Integrals

Now we give some basic definitions that help us define more precisely what we mean by a Riemann integral with respect to product measure in $\mathbf{R}^\infty$. Then we give some conditions for the existence of the Riemann integral with respect to product measure in $\mathbf{R}^\infty$ and go through a certain algorithm useful in computing this integral.

Let $(F_k)_{k \in \mathbf{N}}$ be a sequence of strictly increasing continuous distribution functions on $\mathbf{R}$. Let μ_k be a Borel probability measure in $\mathbf{R}$ defined by F_k for $k \in \mathbf{N}$. Let us denote by $\prod_{k \in \mathbf{N}} \mu_k$ the product of measures $(\mu_k)_{k \in \mathbf{N}}$.

For $-\infty < c < d < +\infty$, let $][c, d][$ denote a subinterval of the real axis $(-\infty, +\infty)$ which has one of the forms $[c, d]$, $[c, d[$, $]c, d[$ or $]c, d]$. If $c = -\infty$ and $d \neq +\infty$, then $][c, d][$ denotes a subinterval of the real axis $(-\infty, +\infty)$ which has one of the forms $]c, d]$ or $]c, d]$. Similarly, if $c \neq -\infty$ and $d = +\infty$, then $][c, d][$ denotes a subinterval of the real axis $(-\infty, +\infty)$ which has one of the forms $]c, d[$ or $[c, d[$. Finally, if $c = -\infty$ and $d = +\infty$, then $][c, d][$ denotes the whole real axis $(-\infty, +\infty)$.

Definition 2.3.1 A set U^* is called an elementary rectangle in $\mathbf{R}^\infty$ if it admits the following representation.

$$U^* = \prod_{k=1}^{m}][c_k, d_k][\times \mathbf{R}^{\mathbf{N} \setminus \{1, \cdots, m\}}, \tag{2.3.1}$$

where $-\infty \leq c_k < d_k \leq +\infty$ for $1 \leq k \leq m$.

Definition 2.3.2 A family of pairwise disjoint elementary rectangles $\tau = (U_k^*)_{1 \leq k \leq n}$ in $\mathbf{R}^\infty$ is called the Riemann partition of the $\mathbf{R}^\infty$ if $\cup_{1 \leq k \leq n} U_k^* = \mathbf{R}^\infty$.

Definition 2.3.3 Let $\tau^* = (U_k^*)_{1 \leq k \leq n}$ be the Riemann partition of $\mathbf{R}^\infty$ and $\ell_1(F_i^{-1}(Pr_i(U_k^*)))$ the length of the preimage of the ith projection $Pr_i(U_k^*)$ of the U_k^* under mapping F_i for $i \in N$. We set

$$d^*(U_k^*) = \sum_{i \in N} \frac{\ell_1(F_i(Pr_i(U_k^*)))}{2^i(1 + \ell_1(F_i(Pr_i(U_k^*))))}. \tag{2.3.2}$$

It is obvious that $d^*(U_k^*)$ is a diameter of the elementary rectangle U_k^* for $k \in N$ with respect to metric ρ defined as

$$\rho((x_k)_{k \in N}, (y_k)_{k \in N}) = \sum_{k \in N} \frac{|F_k(x_k) - F_k(y_k)|}{2^k(1 + |F_k(x_k) - F_k(y_k)|)} \tag{2.3.3}$$

for $(x_k)_{k \in N}, (y_k)_{k \in N} \in \mathbf{R}^\infty$.

Remark 2.3.1 Note that metrics ρ and ρ_T are equivalent provided that

$$\rho((x_k)_{k \in N}, (y_k)_{k \in N}) = 0$$

if and only if

$$\rho_T((x_k)_{k \in N}, (y_k)_{k \in N}) = 0.$$

Note also that both topologies induced by these metrics coincide.

Definition 2.3.4 A number $d^*(\tau)$, defined by

$$d^*(\tau) = \max\{d^*(U_k) : 1 \leq k \leq n\} \tag{2.3.4}$$

is called the mesh or norm of the Riemann partition τ^* of the $\mathbf{R}^\infty$.

Definition 2.3.5 Let $\tau_1^* = (U_i^{*(1)})_{1 \leq i \leq n}$ and $\tau_2^* = (U_j^{*(2)})_{1 \leq j \leq m}$ be Riemann partitions of the $\mathbf{R}^\infty$. We say that $\tau_2^* \leq \tau_1^*$ iff

$$(\forall j)((1 \leq j \leq m) \rightarrow (\exists i_0)(1 \leq i_0 \leq n \,\&\, U_j^{*(2)} \subseteq U_{i_0}^{*(1)})). \tag{2.3.5}$$

Definition 2.3.6 A function f is called a step function on $\mathbf{R}^\infty$ if it can be written as

$$f(x) = \sum_{k=1}^{n} c_k \chi_{U_k^*}(x), \tag{2.3.6}$$

where $\tau^* = (U_k^*)_{1 \leq k \leq n}$ is any Riemann partition of the $\mathbf{R}^\infty$, $c_k \in R$ for $1 \leq k \leq n$ and χ_A is the indicator function of the set A.

Definition 2.3.7 Let f be a real-valued bounded function defined on $\mathbf{R}^\infty$. Let $\tau^* = (U_k^*)_{1 \leq k \leq n}$ be the Riemann partition of $\mathbf{R}^\infty$ and $(t_k^*)_{1 \leq k \leq n}$ be a sample such that, for each k, $t_k^* \in U_k^*$. Then

(i) A sum $\sum_{k=1}^{n} f(t_k^*)(\prod_{i \in \mathbf{N}} \mu_i)(U_k^*)$ is called the Riemann sum of the f with respect to the Riemann partition $\tau^* = (U_k^*)_{1 \leq k \leq n}$ together with sample $(t_k)_{1 \leq k \leq n}$.

(ii) A sum $S_{\tau^*} = \sum_{k=1}^{n} M_k(\prod_{i \in \mathbf{N}} \mu_i)(U_k^*)$ is called the upper Darboux sum with respect to the Riemann partition τ^*, where $M_k = \sup_{x \in U_k^*} f(x)(1 \leq k \leq n)$.

(iii) A sum $s_{\tau^*} = \sum_{k=1}^{n} m_k(\prod_{i \in \mathbf{N}} \mu_i)(U_k^*)$ is called the lower Darboux sum with respect to the Riemann partition τ^*, where $m_k = \inf_{x \in U_k^*} f(x)(1 \leq k \leq n)$.

Definition 2.3.8 Let f be a real-valued bounded function defined on $\mathbf{R}^\infty$. We say that the f is Riemann integrable with respect to product measure $\prod_{i \in \mathbf{N}} \mu_i$ on $\mathbf{R}^\infty$ if there exists a real number s such that for every positive real number ε there exists a real number $\delta > 0$ such that, for every Riemann partition $(U_k^*)_{1 \leq k \leq n}$ of the $\mathbf{R}^\infty$ with $d^*(\tau^*) < \delta$ and for every sample $(t_k^*)_{1 \leq k \leq n}$, we have

$$\left|\sum_{k=1}^{n} f(t_k^*)\left(\prod_{i \in \mathbf{N}} \mu_i\right)(U_k^*) - s\right| < \varepsilon. \tag{2.3.7}$$

The number s is called the Riemann integral of f over $\mathbf{R}^\infty$ and is denoted by

$$(R) \int_{\mathbf{R}^\infty} f(x) d\left(\prod_{i \in \mathbf{N}} \mu_i\right)(x). \tag{2.3.8}$$

In this section we present some conditions that help us determine whether the Riemann integral of a certain function over $\mathbf{R}^\infty$ exists.

Theorem 2.3.1 *(Riemann necessary and sufficient condition for integrability). Consider the bounded function* $f : R^\infty \to R$. *f is Riemann integrable in* $\mathbf{R}^\infty$ *with respect to product measure* $\prod_{i\in\mathbf{N}} \mu_i$ *if and only if for arbitrary positive* ε *there is a Riemann partition* τ^* *of* $\mathbf{R}^\infty$ *such that* $S_{\tau^*} - s_{\tau^*} < \varepsilon$.

The proof of Theorem 2.3.1 can be obtained by the standard scheme.

Example 2.3.1 Define $u((x_k)_{k\in\mathbf{N}}) = sin(x_1^{-1})$ for $(x_k)_{k\in N} \in (0,1)^\infty$. Then u is bounded (by 1) and continuous on $(0,1)^\infty$, but it is neither uniformly continuous nor continuously extendable to $[0,1]^\infty$.

In the context of Example 2.3.1 the following lemma is of some interest.

Lemma 2.3.1 *Let f be any bounded and uniformly continuous function on* $(0,1)^\infty$. *Then f has a unique continuous extension* $\overline{f}$ *to whole* $[0,1]^\infty$.

Proof For any $x \in [0,1]^\infty$, find a sequence $(x_n) \in (0,1)^\infty$ such that $\lim_{n\to\infty} x_n = x$.

Step 1. Because $(x_n)_{n\in\mathbf{N}}$ is Cauchy, and f is uniformly continuous, we deduce that $(f(x_n))_{n\in\mathbf{N}}$ is Cauchy.

Assume the contrary and that $(f(x_n))_{n\in\mathbf{N}}$ is not a Cauchy sequence. Then for some $\varepsilon > 0$ and for each natural number m there are two natural numbers $n_1^{(m)} > m$ and $n_2^{(m)} > m$ such that $|f(x_{n_1^{(m)}}) - f(x_{n_2^{(m)}})| > \varepsilon$.

Let us consider a set $\{x_{n_1^{(m)}}, x_{n_2^{(m)}} : m \in \mathbf{N}\}$.

Because f is a uniformly continuous function on $(0,1)^\infty$, for $\varepsilon/2$ there exists $\delta > 0$ such that if $x, y \in (0,1)^\infty$ and $\rho_T(x,y) < \varepsilon/2$ then $|f(x) - f(y)| < \varepsilon/2$. Inasmuch as $(x_n)_{n\in\mathbf{N}}$ is a Cauchy sequence we can choose such $m \in \mathbf{N}$ that $\rho_T(x_{n_1^{(m_0)}}, x_{n_2^{(m_0)}}) < \delta$. But $|f(x_{n_1^{(m_0)}}) - f(x_{n_2^{(m_0)}})| > \varepsilon$ and we get the contradiction.

Step 2. Define $\overline{f}(x) = \lim_{n\to\infty} f(x_n)$.

Step 3. Let us show that this definition is independent of the choice of the sequence $(x_n)_{n\in\mathbf{N}}$.

Indeed, we have another sequence $(y_n)_{n\in\mathbf{N}}$ of elements of $(0,1)^\infty$ which tends to x. Let us show that $\lim_{n\to\infty} f(y_n) = f(x)$. For $\varepsilon > 0$ there is $n(\varepsilon)$ such that for each $n \geq n(\varepsilon)$ we get $|f(x_n) - f(x)| < \varepsilon/2$.

Because f is uniformly continuous on $(0,1)^\infty$ for $\varepsilon/2$ there is $\delta(\varepsilon, f) > 0$ such that if $\rho_T(w,z) < \delta(\varepsilon, f)$ then $|f(w) - f(z)| < \varepsilon/2$. Because $(y_n)_{n\in\mathbf{N}}$ and $(x_n)_{n\in\mathbf{N}}$ tend to x, for $\delta(\varepsilon, f)/2$ there exists a natural number $n(\delta(\varepsilon, f))$ such that $\rho_T(y_n, x) < \delta(\varepsilon, f)/2$ and $\rho_T(x_n, x) < \delta(\varepsilon, f)/2$ for $n \geq n(\delta(\varepsilon, f))$. Then for $n \geq n(\delta(\varepsilon, f))$ we get

$$\rho_T(x_n, y_n) \leq \rho_T(x_n, x) + \rho_T(x, y_n) < \delta(\varepsilon, f)/2 + \delta(\varepsilon, f)/2 = \delta(\varepsilon, f) \quad (2.3.9)$$

which implies $|f(x_n) - f(y_n)| < \varepsilon/2$.

Then for $n \geq \max\{n(\varepsilon), n(\delta(\varepsilon, f))\}$ we get

$$|f(x) - f(y_n)| = |f(x) - f(x_n) + f(x_n) - f(y_n)| \leq$$

$$|f(x) - f(x_n)| + |f(x_n) - f(y_n)| \leq \varepsilon/2 + \varepsilon/2 = \varepsilon. \tag{2.3.10}$$

Note that $\overline{f}$ is an extension of f (i.e., it coincides with f on $(0,1)^\infty$) because of Step 3.

Uniqueness holds because any continuous extension of f must satisfy the equality of Step 2; that is, if g is another continuous extension of f, then for any $(x_n)_{n\in\mathbf{N}}$ as above $g(x) = \lim_{n\to\infty} f(x_n) = \overline{f}(x)$. As for boundedness, it again follows from Step 2. If $|f(y)| \leq M$ for all $y \in (0,1)^\infty$, then $|\overline{f}(x)| = \lim_{n\to\infty} |f(x_n)| \leq M$ as well.

Let $f : \mathbf{R}^\infty \to \mathbf{R}$ be a real-valued function. We set $f_{(F_i)_{i\in\mathbf{N}}} : (0,1)^\infty \to \mathbf{R}$ as follows. $f_{(F_i)_{i\in\mathbf{N}}}((x_k)_{k\in\mathbf{N}}) = f((F_k^{-1}(x_k))_{k\in\mathbf{N}})$ if $(x_k)_{k\in\mathbf{N}} \in (0,1)^\infty$.

Now it is not hard to prove the following assertion.

Theorem 2.3.2 *Let f be a real-valued bounded function on $\mathbf{R}^\infty$ such that $f_{(F_i)_{i\in\mathbf{N}}}$ admits the Riemann integrable (with respect to the infinite-dimensional "Lebesgue measure" in $[0,1]^\infty$) extension $\overline{f}_{(F_i)_{i\in N}}$ from $(0,1)^\infty$ to whole $[0,1]^\infty$. Then f is Riemann integrable w.r.t. the product measure $\prod_{i\in\mathbf{N}} \mu_i$ and the following equality,*

$$(R)\int_{\mathbf{R}^\infty} f(x)d\left(\prod_{i\in\mathbf{N}} \mu_i\right)(x) = (R)\int_{[0,1]^\infty} \overline{f}_{(F_i)_{i\in\mathbf{N}}}(x)d\lambda(x), \tag{2.3.11}$$

holds true.

Theorem 2.3.3 *If f is a real-valued bounded uniformly continuous function on $\mathbf{R}^\infty$ then f is Riemann integrable w.r.t. the product measure $\prod_{i\in\mathbf{N}} \mu_i$ and the following equality,*

$$(R)\int_{\mathbf{R}^\infty} f(x)d\left(\prod_{i\in\mathbf{N}} \mu_i\right)(x) = (R)\int_{[0,1]^\infty} \overline{f}_{(F_i)_{i\in\mathbf{N}}}(x)d\lambda(x), \tag{2.3.12}$$

holds true, where $\overline{f}_{(F_i)_{i\in\mathbf{N}}}$ is a continuous extension of $f_{(F_i)_{i\in\mathbf{N}}}$ from $(0,1)^\infty$ to whole $[0,1]^\infty$ defined by Lemma 2.3.1.

Proof Because f is bounded and uniformly continuous on $\mathbf{R}^\infty$ with respect to metric ρ we claim that $f_{(F_i)_{i\in\mathbf{N}}}$ is bounded and uniformly continuous on $(0,1)^\infty$ with respect to metric ρ_T. By Lemma 2.3.1, we know that $f_{(F_i)_{i\in\mathbf{N}}}$ has a unique bounded continuous extension $\overline{f}_{(F_i)_{i\in\mathbf{N}}}$ on $[0,1]^\infty$. By Theorem 2.3.2 we know that $\overline{f}_{(F_i)_{i\in\mathbf{N}}}$ is Riemann integrable on $[0,1]^\infty$ w.r.t. λ. This means that there exists a real number s such that for every positive real number ε there exists a real number $\delta > 0$ such that for every

Riemann partition $(U_k)_{1\le k\le n}$ of the $[0,1]^\infty$ with $d(\tau) < \delta$ and for every sample $(t_k)_{1\le k\le n}$, we have

$$|\sum_{k=1}^{n} \overline{f}_{(F_i)_{i\in\mathbf{N}}}(t_k)\lambda(U_k) - s| < \varepsilon. \tag{2.3.13}$$

The latter relation implies that for every Riemann partition $(U_k)_{1\le k\le n}$ of the $[0,1]^\infty$ with $d(\tau) < \delta$ and for every sample $(t_k)_{1\le k\le n}$ for which $t_k \in U_k \cap (0,1)^\infty (1 \le k \le n)$, we have

$$|\sum_{k=1}^{n} f_{(F_i)_{i\in\mathbf{N}}}(t_k)\lambda(U_k \cap (0,1)^\infty) - s| < \varepsilon. \tag{2.3.14}$$

We have to show that s is a real number such that for every positive real number ε, δ is a number such that for every Riemann partition $\tau^* = (U_k^*)_{1\le k\le n}$ of the $\mathbf{R}^\infty$ with $d^*(\tau^*) < \delta$ and for every sample $(t_k^*)_{1\le k\le n}$ with $t_k^* \in U_k^* (1 \le k \le n)$, we have

$$|\sum_{k=1}^{n} f(t_k^*)\left(\prod_{i\in\mathbf{N}} \mu_i\right)(U_k^*) - s| < \varepsilon. \tag{2.3.15}$$

We set $\mathbf{F}((x_k)_{k\in\mathbf{N}}) = (F_k(x_k))_{k\in\mathbf{N}}$ for $(x_k)_{k\in\mathbf{N}} \in \mathbf{R}^\infty$.

If $(U_k^*)_{1\le k\le n}$ is a Riemann partition of $\mathbf{R}^\infty$ with $d^*(\tau^*) < \delta$, then $\tau = (U_k)_{1\le k\le n} := (\mathbf{F}(U_k^*))_{1\le k\le n}$ will be a Riemann partition of $(0,1)^\infty$ with $d(\tau) < \delta$ and $(t_k)_{1\le k\le n} = (\mathbf{F}(^*t_k))_{1\le k\le n}$ will sample from the partition τ such that

$$|\sum_{k=1}^{n} f(t_k^* \times \left(\prod_{i\in\mathbf{N}} \mu_i\right)(U_k^*)) - s| = |\sum_{k=1}^{n} f_{(F_i)_{i\in\mathbf{N}}}(t_k)\lambda(U_k) - s| < \varepsilon. \tag{2.3.16}$$

The latter relation means that

$$(R)\int_{\mathbf{R}^\infty} f(x)d\left(\prod_{i\in\mathbf{N}} \mu_i\right)(x) = s. \tag{2.3.17}$$

On the other hand we have that

$$(R)\int_{[0,1]^\infty} \overline{f}_{(F_i)_{i\in\mathbf{N}}}(x)d\lambda(x) = s. \tag{2.3.18}$$

This ends the proof of the theorem.

The following corollary shows us how the Riemann integral can be computed with respect to the product measure in $\mathbf{R}^\infty$.

Corollary 2.3.1 *Let f be a bounded uniformly continuous real-valued function on $\mathbf{R}^\infty$. Let $(Y_n)_{n\in N}$ be an increasing family of uniformly distributed finite subsets in $[0,1]^\infty$. Then the equality*

$$(R)\int_{\mathbf{R}^\infty} f(x)d\left(\prod_{i\in\mathbf{N}}\mu_i\right)(x) = \lim_{n\to\infty}\frac{\sum_{y\in Y_n}\overline{f}_{(F_i)_{i\in\mathbf{N}}}(y)}{\#(Y_n)} \tag{2.3.19}$$

holds true.

Proof By Theorem 2.3.3 we know that

$$(R)\int_{\mathbf{R}^\infty} f(x)d\left(\prod_{i\in\mathbf{N}}\mu_i\right)(x) = (R)\int_{[0,1]^\infty}\overline{f}_{(F_i)_{i\in\mathbf{N}}}(x)d\lambda(x). \tag{2.3.20}$$

By Theorem 2.2.5 we have

$$(R)\int_{[0,1]^\infty}\overline{f}_{(F_i)_{i\in\mathbf{N}}}(x)d\lambda(x) = \lim_{n\to\infty}\frac{\sum_{y\in Y_n}\overline{f}_{(F_i)_{i\in\mathbf{N}}}(y)}{\#(Y_n)}. \tag{2.3.21}$$

This ends the proof of the corollary.

Remark 2.3.2 Let f be a bounded uniformly continuous real-valued function on $\mathbf{R}^\infty$. It is not hard to show that there is an increasing family of uniformly distributed finite subsets $(Y_n)_{n\in N}$ in $[0,1]^\infty$ such that $Y_n \subseteq (0,1)^\infty$ for each $n \in N$. Then the equality

$$(R)\int_{\mathbf{R}^\infty} f(x)d\left(\prod_{i\in\mathbf{N}}\mu_i\right)(x) = \lim_{n\to\infty}\frac{\sum_{(y_i)_{i\in N}\in Y_n} f((F_i^{-1}(y_i))_{i\in\mathbf{N}})}{\#(Y_n)} \tag{2.3.22}$$

holds true.

The following example can be considered as a certain application of the Remark 2.3.2 in mathematical analysis.

Example 2.3.2 The following equality

$$\lim_{n\to\infty}\frac{\sum_{(i_1,i_2,\cdots,i_n)\in\{1,\cdots,n\}^n}\sum_{k=1}^{n}\frac{\{i_k\omega\}^a}{2^k}}{n^n} = \frac{1}{1+a} \tag{2.3.23}$$

holds true for all irrational numbers ω and positive real numbers α.

Let $f : \mathbf{R}^\infty \to \mathbf{R}$ be defined by $f((x_k)_{k\in\mathbf{N}}) = \sum_{k\in\mathbf{N}} F_k^\alpha(x_k)/2^k$, where $\alpha > 0$. Then

$$f((F_k^{-1}(y_k))_{k\in\mathbf{N}}) = \sum_{k\in\mathbf{N}} \frac{F_k^\alpha(F_k^{-1}(y_k))}{2^k} = \sum_{k\in\mathbf{N}} \frac{y_k^\alpha}{2^k} \tag{2.3.24}$$

for $(y_k)_{k\in\mathbf{N}} \in (0,1)^\infty$.

Let ω be an arbitrary irrational number. Let $Y_n = \{\{\omega\}, \{2\omega\}, \dots, \{n\omega\}\}^n \times (\{\omega\}, \{\omega\}, \dots)$ for $n \in \mathbf{N}$. Then by virtue of Remark 2.3.2 we have

$$(R)\int_{\mathbf{R}^\infty} f(x)d\left(\prod_{i\in\mathbf{N}}\mu_i\right)(x) = \lim_{n\to\infty} \frac{\sum_{(y_i)_{i\in N}\in Y_n} f((F_i^{-1}(y_i))_{i\in\mathbf{N}})}{\#(Y_n)} =$$

$$\lim_{n\to\infty} \frac{\sum_{(i_1,i_2,\cdots,i_n)\in\{1,\cdots,n\}^n}(\sum_{k=1}^n \frac{\{i_k\omega\}^\alpha}{2^k} + \sum_{k>n}\frac{\{\omega\}^\alpha}{2^k})}{n^n} =$$

$$\lim_{n\to\infty} \frac{\sum_{(i_1,i_2,\cdots,i_n)\in\{1,\cdots,n\}^n}\sum_{k=1}^n \frac{\{i_k\omega\}^\alpha}{2^k}}{n^n}. \tag{2.3.25}$$

On the other hand we have

$$(R)\int_{\mathbf{R}^\infty} f(x)d\left(\prod_{i\in\mathbf{N}}\mu_i\right)(x) = (R)\int_{[0,1]^\infty} \sum_{k\in\mathbf{N}} \frac{x_k^\alpha}{2^k} d\lambda(x) =$$

$$\sum_{k\in\mathbf{N}} \frac{1}{2^k}(R)\int_{[0,1]^\infty} x_k^\alpha d\lambda(x) = \frac{1}{1+\alpha}\sum_{k\in\mathbf{N}}\frac{1}{2^k} = \frac{1}{1+\alpha}. \tag{2.3.26}$$

Finally, we get the identity:

$$\lim_{n\to\infty} \frac{\sum_{(i_1,i_2,\dots,i_n)\in\{1,\dots,n\}^n}\sum_{k=1}^n \frac{\{i_k\omega\}^\alpha}{2^k}}{n^n} = \frac{1}{1+\alpha}. \tag{2.3.27}$$

2.4 Applications to Statistics

In probability theory, there exist several different notions of convergence of random variables. The convergence of sequences of random variables to some limit random variable is an important concept in probability theory. Almost sure convergence is called the *strong law* because random variables that converge strongly (almost surely) are guaranteed to converge weakly (in probability) and in distribution (see, e.g., [Sh], Theorem 2, p. 272). Theorems that establish almost sure convergence of such sequences to some limit random variable are called *strong law type theorems* and they have interesting applications to statistics and stochastic processes. The purpose of the present section is to establish the validity of essentially new and interesting strong law type theorems in an infinite-dimensional case by using Monte Carlo algorithms elaborated in Sect. 2.3.

Theorem 2.4.1 *Let* $(\Omega, \mathbf{F}, P)$ *be a probability space and* $(\xi_k)_{k\in N}$ *be a sequence of independent real-valued random variables uniformly distributed on the interval* $[0, 1]$ *such that* $0 \le \xi_k(\omega) \le 1$. *Let* $f : [0, 1]^\infty \to R$ *be a Riemann integrable real-valued function. Then the equality:*

$$P\left\{\omega : \lim_{n\to\infty} \frac{\sum_{(i_1,i_2,\dots,i_n)\in\{1,\dots,n\}^n} f(\xi_{i_1}(\omega), \xi_{i_2}(\omega), \dots, \xi_{i_n}(\omega), \xi_1(\omega), \xi_1(\omega), \dots)}{n^n} = \int_{[0,1]^\infty} f(x)d\lambda(x)\right\} = 1 \quad (2.4.1)$$

holds true.

Proof Without loss of generality we can assume that

$$(\Omega, \mathbf{F}, P) = ([0, 1]^\infty, \mathbf{B}([0, 1]^\infty), \ell_1^\infty), \quad (2.4.2)$$

where ℓ_1 is the Lebesgue measure in $(0, 1)$ and $\xi_k((\omega_i)_{i\in N}) = \omega_k$ for each $k \in N$ and $(\omega_i)_{i\in N} \in [0, 1]^\infty$. Let S be a set of all uniformly distributed sequences on $(0, 1)$. By Lemma 1.2.4 we know that $\ell_1^N(S) = 1$; equivalently, $\lambda(S) = 1$, where λ denotes the infinite-dimensional "Lebesgue measure". The latter relation means that

$$P\{\omega : (\xi_k(\omega))_{k\in N} \text{ is uniformly distributed on } (0, 1)\} = 1. \quad (2.4.3)$$

We put

$$Y_n(\omega) = (\cup_{j=1}^n \{\xi_j(\omega)\})^n \times (\xi_1(\omega), \xi_1(\omega), \dots) \quad (2.4.4)$$

for each $n \in N$.

Note that if the sequence of real numbers $(\xi_k(\omega))_{k\in N}$ is uniformly distributed in the interval $[0, 1]$ then by Theorem 2.2.1, $(Y_n(\omega))_{n\in N}$ will be uniformly distributed in the rectangle $[0, 1]^\infty$ which according to Theorem 2.2.5 implies that

$$\int_{[0,1]^\infty} f(x)d\lambda(x) =$$

$$\lim_{n\to\infty} \frac{\sum_{(i_1,i_2,\dots,i_n)\in\{1,\dots,n\}^n} f(\xi_{i_1}(\omega),\xi_{i_2}(\omega),\dots,\xi_{i_n}(\omega),\xi_1(\omega),\xi_1(\omega),\dots)}{n^n}. \quad (2.4.5)$$

But the set of all ω points for which the latter equality holds true contains the set S for which $P(S) = 1$.

This ends the proof of the theorem.

As a simple consequence of Theorem 2.4.1, we get the validity of the strong law of large numbers for a sequence of independent real-valued random variables uniformly distributed on the interval [0, 1] as follows.

Corollary 2.4.1 *Let* $(\Omega, \mathbf{F}, P)$ *be a probability space and* $(\xi_k)_{k\in N}$ *be a sequence of independent real-valued random variables uniformly distributed on the interval* [0, 1] *such that* $0 \le \xi_k(\omega) \le 1$. *Then the condition*

$$P\left\{\omega : \lim_{n\to\infty} \frac{\sum_{k=1}^{n} \xi_k(\omega)}{n} = 1/2\right\} = 1 \quad (2.4.6)$$

holds true.

Proof Let $f : [0,1]^\infty \to R$ be defined by $f(x_1, x_2, \dots) = x_1$. By Theorem 2.4.1 we have

$$P\{\omega : \int_{[0,1]^\infty} f(x)d\lambda(x) =$$

$$\lim_{n\to\infty} \frac{\sum_{(i_1,i_2,\dots,i_n)\in\{1,\dots,n\}^n} f(\xi_{i_1}(\omega),\xi_{i_2}(\omega),\dots,\xi_{i_n}(\omega),\xi_1(\omega),\xi_1(\omega),\dots)}{n^n}\} = 1. \quad (2.4.7)$$

Note that

$$\int_{[0,1]^\infty} f(x)d\lambda(x) = 1/2 \quad (2.4.8)$$

and

$$\lim_{n\to\infty} \frac{\sum_{(i_1,i_2,\dots,i_n)\in\{1,\dots,n\}^n} f(\xi_{i_1}(\omega),\xi_{i_2}(\omega),\dots,\xi_{i_n}(\omega),\xi_1(\omega),\xi_1(\omega),\dots)}{n^n} =$$

$$\lim_{n\to\infty} \frac{\sum_{(i_1,i_2,\dots,i_n)\in\{1,\dots,n\}^n} \xi_{i_1}(\omega)}{n^n} =$$

$$\lim_{n\to\infty}\frac{n^{n-1}\sum_{k=1}^{n}\xi_k(\omega)}{n^n}=\lim_{n\to\infty}\frac{\sum_{k=1}^{n}\xi_k(\omega)}{n}. \tag{2.4.9}$$

This ends the proof of Corollary 2.4.1.

The next corollary also being a simple consequence of Theorem 2.4.1 gives interesting but well-known information for statisticians regarding whether the value of m-dimensional Riemann integrals over the m-dimensional rectangle $[0, 1]^m$ can be estimated by using infinite samples.

Corollary 2.4.2 *Let* $(\Omega, \mathbf{F}, P)$ *be a probability space and* $(\xi_k)_{k\in N}$ *be a sequence of independent real-valued random variables uniformly distributed on the interval* $[0, 1]$ *such that* $0 \leq \xi_k(\omega) \leq 1$. *Let* $f : [0, 1]^m \to R$ *be a Riemann integrable real-valued function. Then the equality*

$$P\{\omega : \lim_{n\to\infty}\frac{\sum_{(i_1,i_2,\dots,i_m)\in\{1,\dots,n\}^m} f(\xi_{i_1}(\omega),\xi_{i_2}(\omega),\dots,\xi_{i_m}(\omega))}{n^m}=$$

$$\int_{[0,1]^m} f(x_1,\dots,x_m)dx_1\dots dx_m\}=1 \tag{2.4.10}$$

holds true.

Proof For $(x_k)_{k\in N}\in[0, 1]^\infty$ we put $\overline{f}((x_k)_{k\in N}) = f(x_1,\dots,x_m)$. Without loss of generality we can assume that

$$(\Omega, \mathbf{F}, P) = ([0, 1]^\infty, \mathbf{B}([0, 1]^\infty), \ell_1^\infty), \tag{2.4.11}$$

where ℓ_1 is the Lebesgue measure in $(0, 1)$ and $\xi_k((\omega_i)_{i\in N}) = \omega_k$ for each $k\in N$ and $(\omega_i)_{i\in N}\in[0, 1]^\infty$. Let S be a set of all uniformly distributed sequences on $(0, 1)$. By Lemma 1.2.4 we know that $P(S) = 1$. The latter relation means that

$$P\{\omega : (\xi_k(\omega))_{k\in N} \text{ is uniformly distributed on the interval } (0, 1)\} = 1. \tag{2.4.12}$$

We put

$$Y_n(\omega) = (\cup_{j=1}^{n}\{\xi_j(\omega)\})^m \times (\xi_1(\omega),\xi_1(\omega),\dots) \tag{2.4.13}$$

for each $n\in N$.

Note that if $(\xi_k(\omega))_{k\in N}$ is uniformly distributed on the interval $(0, 1)$ then by Theorem 2.2.1, $(Y_n(\omega))_{n\in N}$ will be uniformly distributed in the rectangle $[0, 1]^\infty$, which according to Theorem 2.2.5 implies that

$$\int_{[0,1]^m} f(x_1,\dots,x_m)dx_1\dots dx_m = \int_{[0,1]^\infty}\overline{f}(x)d\lambda(x) =$$

$$\lim_{n\to\infty}\frac{\sum_{(i_1,i_2,\dots,i_n)\in\{1,\dots,n\}^n}\overline{f}(\xi_{i_1}(\omega),\xi_{i_2}(\omega),\dots,\xi_{i_n}(\omega),\xi_1(\omega),\xi_1(\omega),\dots)}{n^n}=$$

$$\lim_{n\to\infty}\frac{\sum_{(i_1,i_2,\dots,i_n)\in\{1,\dots,n\}^n}f(\xi_{i_1}(\omega),\xi_{i_2}(\omega),\dots,\xi_{i_m}(\omega))}{n^n}=$$

$$\lim_{n\to\infty}\frac{\sum_{(i_1,i_2,\dots,i_m)\in\{1,\dots,n\}^m}n^{n-m}f(\xi_{i_1}(\omega),\xi_{i_2}(\omega),\dots,\xi_{i_m}(\omega))}{n^n}=$$

$$\lim_{n\to\infty}\frac{\sum_{(i_1,i_2,\dots,i_m)\in\{1,\dots,n\}^m}f(\xi_{i_1}(\omega),\xi_{i_2}(\omega),\dots,\xi_{i_m}(\omega))}{n^m}. \tag{2.4.14}$$

A set of all points ω for which the latter equality holds true, contains the set S for which $P(S)=1$.

This ends the proof of Corollary 2.4.2.

Corollary 2.4.3 *Let* $(\Omega,\mathbf{F},P)$ *be a probability space and* $(\xi_k)_{k\in N}$ *be a sequence of independent real-valued random variables such that the distribution function* F_k *defined by* ξ_k *is strictly increasing and continuous. Let* f *be a real-valued bounded function on* $\mathbf{R}^\infty$ *such that* $f_{(F_i)_{i\in\mathbf{N}}}$ *admits such an extension* $\overline{f}_{(F_i)_{i\in N}}$ *from* $(0,1)^\infty$ *to whole* $[0,1]^\infty$ *that* $\overline{f}_{(F_i)_{i\in N}}$ *is Riemann integrable with respect to the infinite-dimensional Lebesgue measure* λ *in* $[0,1]^\infty$*. Then* f *is Riemann integrable w.r.t. product measure* $\prod_{i\in\mathbf{N}}\mu_i$ *and the condition*

$$P\left\{\omega:\lim_{n\to\infty}\frac{\sum_{(i_1,i_2,\dots,i_n)\in\{1,\dots,n\}^n}f(\xi_{i_1}(\omega),\xi_{i_2}(\omega),\dots,\xi_{i_n}(\omega),\xi_1(\omega),\xi_1(\omega),\dots)}{n^n}=\right.$$

$$\left.(R)\int_{\mathbf{R}^\infty}f(x)d\left(\prod_{i\in\mathbf{N}}\mu_i\right)(x)\right\}=1 \tag{2.4.15}$$

holds true.

Proof Without loss of generality we can assume that

$$(\Omega,\mathbf{F},P)=\left(\mathbf{R}^\infty,\mathbf{B}(\mathbf{R}^\infty),\prod_{i\in\mathbf{N}}\mu_i\right), \tag{2.4.16}$$

and $\xi_k((\omega_i)_{i\in N})=\omega_k$ for each $k\in N$ and $(\omega_i)_{i\in N}\in\mathbf{R}^\infty$.

Let ω be an element of the Ω such that $(F_k(\xi_k(\omega))_{k\in N}$ is a uniformly distributed sequence on $(0,1)$. Note that all such points ω constitute a set D_0 for which $(\prod_{i\in\mathbf{N}}\mu_i)(D_0)=1$.

According to Theorem 2.3.2, f is Riemann integrable with respect to the product measure $\prod_{i\in \mathbf{N}} \mu_i$ and the equality

$$(R)\int_{\mathbf{R}^\infty} f(x)d\left(\prod_{i\in \mathbf{N}} \mu_i\right)(x) = (R)\int_{[0,1]^\infty} \overline{f}_{(F_i)_{i\in N}}(x)d\lambda(x) \tag{2.4.17}$$

holds true. For $\omega \in D_0$ we have

$$\begin{aligned}
&(R)\int_{[0,1]^\infty} \overline{f}_{(F_i)_{i\in N}}(x)d\lambda(x)\\
&= \lim_{n\to\infty} \frac{\sum_{(i_1,i_2,\dots,i_n)\in\{1,\dots,n\}^n} \overline{f}_{(F_i)_{i\in N}}(F_1(\xi_{i_1}(\omega)),\dots,F_n(\xi_{i_n}(\omega)),F_{n+1}(\xi_1(\omega)),F_{n+2}(\xi_1(\omega)),\dots)}{n^n}\\
&= \lim_{n\to\infty} \frac{\sum_{(i_1,i_2,\dots,i_n)\in\{1,\dots,n\}^n} f_{(F_i)_{i\in N}}(F_1(\xi_{i_1}(\omega)),\dots,F_n(\xi_{i_n}(\omega)),F_{n+1}(\xi_1(\omega)),F_{n+2}(\xi_1(\omega)),\dots)}{n^n}\\
&= \lim_{n\to\infty} \frac{\sum_{(i_1,i_2,\dots,i_n)\in\{1,\dots,n\}^n} f(F_1^{-1}(F_1(\xi_{i_1}(\omega)),\dots,F_n^{-1}(F_n(\xi_{i_n}(\omega)),F_{n+1}^{-1}(F_{n+1}(\xi_1(\omega)),\dots)}{n^n}\\
&= \lim_{n\to\infty} \frac{\sum_{(i_1,i_2,\dots,i_n)\in\{1,\cdots,n\}^n} f(\xi_{i_1}(\omega),\dots,\xi_{i_n}(\omega),\xi_1(\omega),\xi_1(\omega),\dots)}{n^n}.
\end{aligned} \tag{2.4.18}$$

This ends the proof of Corollary 2.4.3.

Remark 2.4.1 Main results of Sect. 2.3–2.4 were obtained in [P3].

References

[B1] Baker, R.: Lebesgue measure on $\mathbf{R}^\infty$. Proc. Am. Math. Soc. **113**(4), 1023–1029 (1991)

[H1] Hardy, G., Littlewood, J.: Some problems of diophantine approximation. Acta Math. **37**(1), 193–239 (1914)

[H2] Hardy, G., Littlewood, J.: Some problems of diophantine approximation. Acta Math. **37**(1), 155–191 (1914)

[HSY] Hunt, B.R., Sauer, T., Yorke, J.A.: Prevalence: a translation-invariant "Almost Every" on infinite-dimensional spaces. Bull. (New Ser.) Am. Math. Soc. **27**(2), 217–238 (1992)

[KN] Kuipers, L., Niederreiter, H.: Uniform Distribution of Sequences. Wiley-Interscience, New York-London-Sydney (1974)

[N] Nikolski, S.M.: Course of mathematical analysis (in Russian), no. 1, Moscow (1983)

[P2] Pantsulaia, G.R.: On uniformly distributed sequences of an increasing family of finite sets in infinite-dimensional rectangles. Real Anal. Exchange. **36**(2), 325–340 (2010/2011)

[P3] Pantsulaia, G.R.: Infinite-dimensional Monte-Carlo integration. Monte Carlo Methods Appl. **21**(4), 283–299 (2015)

[P4] Pantsulaia, G.: On generators of shy sets on Polish topological vector spaces. New York J. Math. **14**, 235–261 (2008)

[Sh] Shiryaev, A.N.: Problems in probability. Translated by Andrew Lyasoff. Problem Books in Mathematics. Springer, New York (2012)

[Ul] Ulam, S.: Zur Masstheorier in der allgemeinen Mengenlehre. Fund. Math. **16**, 140–150 (1930)

[Ur] Urysohn, P.: Zum Metrisationsproblem. (German). Math. Ann. **94**(1), 309–315 (1925)

[V] Vinogradow, I.: On fractional parts of certain functions. Ann. Math. **37**(2), 448–455 (1936)

[W] Weyl, H.: Úber ein problem aus dem Gebiete der diophantischen approximation. Marchr. Ges. Wiss., Gótingen, Math-phys. K1., 234–244 (1916)

Chapter 3
Structure of All Real-Valued Sequences Uniformly Distributed in $[-\frac{1}{2}, \frac{1}{2}]$ from the Point of View of Shyness

3.1 Introduction

It is well known that nowadays the theory of uniform distribution has many interesting applications in various branches of mathematics, such as number theory, probability theory, mathematical statistics, functional analysis, topological algebra, and so on. Therefore, research on the internal structure of all uniformly distributed sequences has not lost relevance today. For example, in [P5] the set D of all real-valued sequences uniformly distributed in $[-\frac{1}{2}, \frac{1}{2}]$ has been studied in terms of the Moore–Yamasaki–Kharazishvili measure μ (cf. [Y, K]) and it has been demonstrated that μ almost every element of $\mathbf{R}^\infty$ is uniformly distributed in $[-\frac{1}{2}, \frac{1}{2}]$.

The purpose of the present chapter is to study structures of D and F from the point of view of shyness [HSY], where F denotes the set of all sequences uniformly distributed modulo 1 in $[-\frac{1}{2}, \frac{1}{2}]$.

The rest of this chapter is the following.

In Sect. 3.2 we consider some auxiliary notions and facts from mathematical analysis and measure theory. In Sect. 3.3 we prove that D is shy in R^N. In Sect. 3.4, we demonstrate that in the Solovay model [Sol] the set F is the prevalent set [HSY] in R^N.

3.2 Some Auxiliary Notions and Facts from Mathematical Analysis and Measure Theory

Let us consider some notions and auxiliary facts from mathematical analysis and measure theory that are useful for our further investigations.

Definition 3.2.1 A sequence of real numbers $(x_k)_{k\in N} \in \mathbf{R}^\infty$ is called uniformly distributed in $[a, b]$ (abbreviated u.d. in [a,b]) if for each c, d with $a \leq c < d \leq b$ we have

G. Pantsulaia, *Applications of Measure Theory to Statistics*,
DOI 10.1007/978-3-319-45578-5_3

$$\lim_{n\to\infty}\frac{\#(\{x_k : 1\le k\le n\}\cap[c,d])}{n}=\frac{d-c}{b-a}, \qquad (3.2.1)$$

where $\#(\cdot)$ denotes the counter measure of a set.

Let λ be the Lebesgue measure on $[0, 1]$. This measure induces the product measure λ_∞ in $[0, 1]^\infty$.

Lemma 3.2.1 ([KN], Theorem 2.2, p. 183) *Let S be the set of all sequences u.d. in* $[0, 1]$, *viewed as a subset of* $\mathbf{R}^\infty$. *Then* $\lambda_\infty(S\cap[0,1]^\infty)=1$.

Let V be a complete metric linear space, by which we mean a vector space (real or complex) with a complete metric for which the operations of addition and scalar multiplication are continuous. When we speak of a measure on V we always mean a nonnegative measure that is defined on the Borel sets of V and is not identically zero. We write $S+v$ for the translation of a set $S\subseteq V$ by a vector $v\in V$.

Definition 3.2.2 ([HSY], *Definition 1, p. 221*) A measure μ is said to be transverse to a Borel set $S\subset V$ if the following two conditions hold.

1. there exists a compact set $U\subset V$ for which $0<\mu(U)<1$.
2. $\mu(S+v)=0$ for every $v\in V$.

Definition 3.2.3 ([HSY], *Definition 2, p. 222*) A Borel set $S\subset V$ is called shy if there exists a measure transverse to S. More generally, a subset of V is called shy if it is contained in a Borel shy set. The complement of a shy set is called a prevalent set.

Definition 3.2.4 ([P4], *Definition* 2.4, *p.237*) A Borel measure μ in V is called a generator of shy sets Generator of shy sets in V, if

$$(\forall X)(\overline{\mu}(X)=0\to X\in S(V)), \qquad (3.2.2)$$

where $\overline{\mu}$ denotes a usual completion of the Borel measure μ, where $S(V)$ denotes the σ ideal of all shy sets in V.

Lemma 3.2.2 ([P4], Theorem 2.4, p. 241) *Every quasi-finite*[1] *translation-quasi-invariant*[2] *the Borel measure μ defined in a Polish topological vector space V is a generator of shy sets.*

The key ingredient for our investigation is the well-known lemma from probability theory.

[1] A measure μ is called quasi-finite if there exists a μ measurable set A for which $0<\mu(A)<+\infty$.

[2] A Borel measure μ defined in a Polish topological vector space V is called translation-quasi-invariant if for each μ measurable set A and any $h\in V$, the following conditions $\mu(A)=$ and $\mu(A+h)=0$ are equivalent.

Lemma 3.2.3 ([Sh], Borelli–Cantelli lemma, p. 271) *Let $(\Omega, \mathbf{F}, P)$ be a probability space. Let $(E_n)_{n\in N}$ be a sequence of events such that*

$$\sum_{n=1}^{\infty} P(E_n) < \infty. \tag{3.2.3}$$

Then the probability that infinitely many of them occur is 0; that is,

$$P\left(\limsup_{n\to\infty} E_n\right) = 0. \tag{3.2.4}$$

Here, "$\limsup_{n\to\infty}$*" denotes the limit supremum of the sequence of events $(E_n)_{n\in N}$ defined by*

$$\limsup_{n\to\infty} E_n = \bigcap_{n=1}^{\infty}\bigcup_{k=n}^{\infty} E_k. \tag{3.2.5}$$

Below we present a certain example of a translation-invariant Borel measure in the Solovay model (SM) [Sol] which is the following system of axioms.

$$(ZF)\&(DC)\&(\text{every subset of } \mathbf{R} \text{ is measurable in the Lebesgue sense}), \tag{3.2.6}$$

where (ZF) denotes the Zermelo–Fraenkel set theory and (DC) denotes the axiom of dependent choices.

Lemma 3.2.4 ([P3], Corollary 1, p. 64) *(SM) Let J be any nonempty subset of the set of all natural numbers N. Let, for $k \in J$, S_k be the unit circle in the Euclidean plane $\mathbf{R}^2$. We may identify the circle S_k with a compact group of all rotations of $\mathbf{R}^2$ about its origin. Let λ_J be the probability Haar measure defined on the compact group $\prod_{k\in J} S_k$. Then the completion $\overline{\lambda_J}$ of λ_J is defined on the power set of $\prod_{k\in J} S_k$.*

For $k \in N$, define the function f_k by $f_k(x) = \exp\{2\pi x i\}$ for every $x \in \mathbf{R}$. For $E \subset \mathbf{R}^N$ and $g \in \prod_{k\in N} S_k$, put

$$f_E(g) = \begin{cases} \text{card}\left(\left(\prod_{k\in N} f_k\right)^{-1}(g) \cap E\right), & \text{if this is finite;} \\ +\infty, & \text{in all other cases.} \end{cases} \tag{3.2.7}$$

Define the functional μ_N by

$$(\forall E)(E \subset \mathbf{R}^N \to \mu_N(E) = \int_{\prod_{k\in N} S_k} f_E(g) d\overline{\lambda_N}(g)). \tag{3.2.8}$$

Lemma 3.2.5 ([P3], Lemma 3, p. 65) *(SM)* μ_N *is a translation-invariant measure defined on the powerset* $\mathbf{R}^N$ *such that* $\mu_N([0;1]^N) = 1$.

3.3 On the Structure of Real-Valued Sequences Uniformly Distributed in $[-\frac{1}{2}, \frac{1}{2}]$ from the Point of View of Shyness

Lemma 3.3.1 *For* $n \in N$, *we put* $E_n = \{(x_k)_{k\in N} : (x_k)_{k\in N} \in \mathbf{R}^N \ \&\ x_n \in [-1/2, 1/2]\}$. *Then* $D \subseteq \limsup_{n\to\infty} E_n$.

Proof Suppose that for some $(x_k)_{k\in N} \in D$ we have $(x_k)_{k\in N} \notin \limsup_{n\to\infty} E_n$. This means that there is $n_0 \in N$ such that $x_k \notin [-1/2, 1/2]$ for each $k \geq n_0$.

Then we get

$$\begin{aligned} &\lim_{n\to\infty} \frac{\#(\{x_1,\ldots,x_n\} \cap [-1/2, 1/2])}{n} \\ &= \lim_{n\to\infty} \frac{\#(\{x_1,\ldots,x_{n_0}, x_{n_0+1},\ldots,x_n\} \cap [-1/2, 1/2])}{n} \\ &= \lim_{n\to\infty} \frac{\#(\{x_1,\ldots,x_{n_0-1}\} \cap [-1/2, 1/2])}{n} \\ &+ \lim_{n\to\infty} \frac{\#(\{x_{n_0}, x_{n_0+1},\ldots,x_n\} \cap [-1/2, 1/2])}{n} \\ &\leq \lim_{n\to\infty} \frac{n_0 - 1}{n} = 0. \end{aligned} \tag{3.3.1}$$

Because $(x_k)_{k\in N} \in D$, on the other hand, we get

$$\lim_{n\to\infty} \frac{\#(\{x_1,\ldots,x_n\} \cap [-1/2, 1/2])}{n} = 1. \tag{3.3.2}$$

This is the contradiction and Lemma 3.3.1 is proved.

Lemma 3.3.2 $\limsup_{n\to\infty} E_n$ *is a Borel shy set in* $\mathbf{R}^N$.

Proof For $m \in R, \sigma > 0$, we put:

(i) $\xi_{(m,\sigma)}$ is a Gaussian random variable on $\mathbf{R}$ with parameters (m, σ). $\Phi_{(m,\sigma)}$ is a distribution function of $\xi_{(m,\sigma)}$. $\gamma_{(m,\sigma)}$ is a linear Gaussian measure in $\mathbf{R}$ defined by $\Phi_{(m,\sigma)}$.

For $n \in N$, let μ_n be a linear Gaussian measure $\gamma_{(0,\sigma_n)}$ in $\mathbf{R}$ such that

$$\frac{1}{\sqrt{2\pi}\sigma_n} \int_{-1/2}^{-1/2} e^{-\frac{t^2}{2\sigma_n^2}} dt \leq \frac{1}{2^n}. \tag{3.3.3}$$

Such a measure always exists. Indeed, we can take under μ_n such a linear Gaussian measure $\gamma_{(0,\sigma_n)}$ for which $\sigma_n > \frac{2^n}{\sqrt{2\pi}}$ for each $n \in N$.

Let us show that the product-measure $\prod_{n\in N}\mu_n$ is a transverse to $\limsup_{n\to\infty} E_n$. We have to show that

$$\left(\prod_{n\in N}\mu_n\right)\left(\limsup_{n\to\infty} E_n + (h_n)_{n\in N}\right) = 0 \tag{3.3.4}$$

for each $(h_n)_{n\in N} \in \mathbf{R}^N$. Note that

$$\limsup_{n\to\infty} E_n + (h_n)_{n\in N} = \limsup_{n\to\infty} E_n^{(h_n)}, \tag{3.3.5}$$

where

$$E_n^{(h_n)} = \left\{(x_k)_{k\in N} : x_n \in \left[-\frac{1}{2} + h_n, \frac{1}{2} + h_n\right] \ \&\ x_k \in \mathbf{R} \text{ for } k \in N \setminus \{n\}\right\}. \tag{3.3.6}$$

Note that

$$\left(\prod_{n\in N}\mu_n\right)(E_n^{(h_n)}) = \mu_n\left(\left[-\frac{1}{2} + h_n, \frac{1}{2} + h_n\right]\right) \le \mu_n\left(\left[-\frac{1}{2}, \frac{1}{2}\right]\right) \le \frac{1}{2^n}. \tag{3.3.7}$$

The latter relation guarantees that

$$\sum_{n=1}^{\infty}\left(\prod_{n\in N}\mu_n\right)(E_n^{(h_n)}) < \infty, \tag{3.3.8}$$

which according to the Borelli–Cantelli lemma (cf. Lemma 3.2.3) implies that

$$\left(\prod_{n\in N}\mu_n\right)(\limsup_{n\to\infty} E_n^{(h_n)}) = 0. \tag{3.3.9}$$

Inasmuch as E_n is Borel measurable in R^N for each $n \in N$, we deduce that $\limsup_{n\to\infty} E_n$ is also Borel measurable. Finally we claim that $\limsup_{n\to\infty} E_n$ is a Borel shy set in $\mathbf{R}^N$ because $\prod_{n\in N}\mu_n$ is the measure transverse to $\limsup_{n\to\infty} E_n$.

This ends the proof of Lemma 3.3.2.

The next proposition is a simple consequence of Lemmas 3.3.1 and 3.3.2.

Theorem 3.3.1 *D is a Borel shy set in* $\mathbf{R}^N$.

Let S come from Lemma 3.2.1. Because $S = D + (1/2, 1/2, \dots)$ and the property of shyness is preserved under translations, by using Theorem 3.3.1 we get the following corollary.

Corollary 3.3.1 *S is a Borel shy set in* $\mathbf{R}^N$.

3.4 On the Structure of Real-Valued Sequences Uniformly Distributed Modulo 1 in $[-\frac{1}{2}, \frac{1}{2}]$ from the Point of View of Shyness

Definition 3.4.1 A sequence of real numbers $(x_k)_{k\in N} \in \mathbf{R}^\infty$ is said to be uniformly distributed modulo 1 (abbreviated u.d. mod 1) if for each c, d with $0 \le c < d \le 1$ we have

$$\lim_{n\to\infty} \frac{\#(\{x_k : 1 \le k \le n\} \cap [c, d])}{n} = d - c. \tag{3.4.1}$$

The set of all sequences u.d. mod 1 is denoted by E.

Definition 3.4.2 A sequence of real numbers $(x_k)_{k\in N} \in \mathbf{R}^\infty$ is said to be uniformly distributed modulo 1 in $[-1/2, 1/2]$ if for each c, d with $-1/2 \le c < d \le 1/2$ we have

$$\lim_{n\to\infty} \frac{\#(\{x_k : 1 \le k \le n\} \cap [c, d])}{n} = d - c. \tag{3.4.2}$$

The set of all sequences u.d. mod 1 in $[-1/2, 1/2]$ is denoted by F.

Remark 3.4.1 Note that $(x_k)_{k\in N}$ is uniformly distributed modulo 1 if and only if $(x_k)_{k\in N}$ is uniformly distributed modulo 1 in $[-1/2, 1/2]$; that is, $E = F$.

In the sequel we need the following lemma.

Lemma 3.4.1 ([KN], Theorem 1.1, p. 2) *The sequence $(x_n)_{n\in N}$ of real numbers is u.d. mod 1 if and only if for every real-valued continuous function f defined on the closed unit interval $\overline{I} = [0, 1]$ we have*

$$\lim_{N\to\infty} \frac{\sum_{n=1}^{N} f(\{x_n\})}{N} = \int_{\overline{I}} f(x)dx. \tag{3.4.3}$$

Theorem 3.4.1 (SM) *The set E of all real-valued sequences uniformly distributed modulo 1 is the prevalent set in* $\mathbf{R}^\infty$.

Proof Let E_0 be the set of all sequences from $(0, 1)^\infty$ that are not uniformly distributed in $[0, 1]$. Because the measure λ_∞ from Lemma 3.2.1 and the measure μ_N from

Lemma 3.2.5 coincide on subsets of $(0, 1)^\infty$ in the Solovay model, by Lemma 3.2.1 we deduce that $\mu_N(E_0) = 0$.

By the definition of the functional μ_N we have

$$\mu_N(E_0) = \int_{\prod_{k \in N} S_k} f_{E_0}(g) d\overline{\lambda_N}(g)) = 0. \tag{3.4.4}$$

We put

$$X_n = \left\{ g : g \in \prod_{k \in N} S_k \ \& \ \mathrm{card}\left(\prod_{k \in N} f_k \right)^{-1} (g) \cap E_0) = n \right\} \tag{3.4.5}$$

for $n \in N \cup \{+\infty\}$. Then we get

$$f_{E_0}(g) = \sum_{n \in N \cup \{+\infty\}} n \chi_{X_n}(g). \tag{3.4.6}$$

Because

$$\mu_N(E_0) = \sum_{n \in N \cup \{+\infty\}} n \overline{\lambda_N}(X_n) = 0 \tag{3.4.7}$$

and

$$\mathrm{card}\left(\left(\prod_{k \in N} f_k \right)^{-1} (g) \cap E_0 \right) \leq 1 \tag{3.4.8}$$

for each $g \in \prod_{k \in N} S_k$, we claim that

$$\overline{\lambda_N}(X_n) = 0 \tag{3.4.9}$$

for each $n \in (N \setminus \{0\}) \cup \{+\infty\}$, which implies that

$$\overline{\lambda_N}(X_0) = 1. \tag{3.4.10}$$

Now let E^* be the set of all sequences of real numbers that are not uniformly distributed modulo 1. Then we get

$$f_{E^*}(g) = \sum_{n \in N \cup \{+\infty\}} n\chi_{Y_n}(g) \tag{3.4.11}$$

where

$$Y_n = \left\{ g : g \in \prod_{k \in N} S_k \ \& \ \mathrm{card}\left(\left(\prod_{k \in N} f_k \right)^{-1} (g) \cap E^* \right) = n \right\}. \tag{3.4.12}$$

Let us show that $X_0 \subseteq Y_0$. Assume the contrary. Then for some $g \in X_0$ and $n > 0$ we get

$$0 = \mathrm{card}\left(\left(\prod_{k \in N} f_k \right)^{-1} (g) \cap E_0 \right) < \mathrm{card}\left(\left(\prod_{k \in N} f_k \right)^{-1} (g) \cap E^* \right) = n, \tag{3.4.13}$$

which implies an existence of such a sequence $(x_k)_{k \in N} \in (\prod_{k \in N} f_k)^{-1}(g) \cap E^*$ for which

$$(\{x_k\})_{k \in N} \in \left(\prod_{k \in N} f_k \right)^{-1} (g) \cap E^*. \tag{3.4.14}$$

Then we also get that

$$(\{x_k\})_{k \in N} \in \left(\prod_{k \in N} f_k \right)^{-1} (g) \cap E_0 \tag{3.4.15}$$

which is the contradiction and we proved that $X_0 \subseteq Y_0$.

Inasmuch as $X_0 \subseteq Y_0$ and $\overline{\lambda_N}(X_0) = 1$, we claim that $\overline{\lambda_N}(Y_0) = 1$. The latter relation implies that $\overline{\lambda_N}(Y_n) = 0$ for each $n \in (N \setminus \{0\}) \cup \{+\infty\}$. Finally we get

$$\mu_N(E^*) = \sum_{n \in N \cup \{+\infty\}} n\overline{\lambda_N}(Y_n) = 0. \tag{3.4.16}$$

Because μ_N is the completion of a quasi-finite translation-invariant Borel measure in R^N, by Lemma 3.2.2 we easily deduce that μ_N is the generator of shy sets Generator of shy sets in R^N, which implies that E^* is shy. The latter relation implies that the set $E = R^N \setminus E^*$ is the prevalent set in $\mathbf{R}^\infty$.

This ends the proof of the theorem.

By using Remark 3.4.1, we get the following corollary of Theorem 3.4.1.

Corollary 3.4.1 (SM) *The set F of all real-valued sequences, uniformly distributed modulo 1 in* $[-1/2, 1/2]$, *is the prevalent set in* $\mathbf{R}^\infty$.

By using Lemma 3.4.1 and Theorem 3.4.1 we get the following versions of the strong law of large numbers in terms of the prevalent set.

Corollary 3.4.2 (SM) *Let f be a real-valued continuous function f defined on the closed unit interval $\overline{I} = [0, 1]$. Then*

$$\left\{(x_n)_{n\in N} \in R^N : \lim_{N\to\infty} \frac{\sum_{n=1}^{N} f(\{x_n\})}{N} = \int_{\overline{I}} f(x)dx\right\} \tag{3.4.17}$$

is the prevalent set in R^N.

Corollary 3.4.3 (SM) *The set*

$$\cap_{f\in C[0,1]} \left\{(x_n)_{n\in N} \in R^N : \lim_{N\to\infty} \frac{\sum_{n=1}^{N} f(\{x_n\})}{N} = \int_{\overline{I}} f(x)dx\right\} \tag{3.4.18}$$

is the prevalent set in R^N.

Proof By Lemma 3.4.1, we know that

$$E \subseteq \left\{(x_n)_{n\in N} \in R^N : \lim_{N\to\infty} \frac{\sum_{n=1}^{N} f(\{x_n\})}{N} = \int_{\overline{I}} f(x)dx\right\} \tag{3.4.19}$$

for each $f \in C[0, 1]$. The latter relation implies that the following inclusion

$$E \subseteq \cap_{f\in C[0,1]} \left\{(x_n)_{n\in N} \in R^N : \lim_{N\to\infty} \frac{\sum_{n=1}^{N} f(\{x_n\})}{N} = \int_{\overline{I}} f(x)dx\right\} \tag{3.4.20}$$

holds true.

Application of the result of Theorem 3.4.1 ends the proof of the corollary.

Remark 3.4.2 Main results of Sect. 3.4 were obtained in [P6]

References

[HSY] Hunt, B.R., Sauer, T., Yorke, J.A.: Prevalence: a translation-invariant "Almost Every" on infinite-dimensional spaces. Bull. (New Ser.) Am. Math. Soc. **27**(2), 217–238 (1992)

[K] Kharazishvili, A.B.: On invariant measures in the Hilbert space (in Russian). Bull. Acad. Sci. Georgian SSR. **114**(1), 41–48 (1984)

[KN] Kuipers, L., Niederreiter, H.: Uniform Distribution of Sequences. Wiley, London (1974)

[P3] Pantsulaia, G.R.: Relations between shy sets and sets of ν_p-measure zero in Solovay's model. Bull. Polish Acad. Sci. **52**(1), 63–69 (2004)

[P4] Pantsulaia, G.R.: On generators of shy sets on Polish topological vector spaces. New York J. Math. **14**, 235–261 (2008)

[P5] Pantsulaia, G.R.: On uniformly distributed sequences on $[-1/2, 1/2]$. Georg. Int. J. Sci. Tech. **4**(3), 21–27 (2013)

[P6] Pantsulaia, G.R.: On structure of uniformly distributed sequences in $[-\frac{1}{2}, \frac{1}{2}]$ from the point of view of shyness. http://arxiv.org/pdf/1506.07009v2.pdf

[Sh] Shiryaev, A.N.: Problems in probability, Translated by Lyasoff, A. Problem Books in Mathematics. Springer, New York (2012)

[Sol] Solovay, R.M.: A model of set theory in which every set of reals is Lebesgue measurable. Ann. Math. **92**, 1–56 (1970)

[Y] Yamasaki, Y.: Translationally invariant measure on the infinite-dimensional vector space. Publ. Res. Inst. Math. Sci. **16**(3), 693–720 (1980)

Chapter 4
On Moore–Yamasaki–Kharazishvili Type Measures and the Infinite Powers of Borel Diffused Probability Measures on R

4.1 Introduction

Let μ and ν be non-trivial σ-finite measures on a measurable space (X, M). The measures μ and ν are called orthogonal if there is a measurable set $E \in M$ such that $\mu(E) = 0$ and $\nu(X \setminus E) = 0$. The measures μ and ν are called equivalent if and only if the following condition

$$(\forall E)(E \in M \rightarrow (\mu(E) = 0 \Longleftrightarrow \nu(E) = 0)) \tag{4.1.1}$$

holds true.

It is well known that the following facts are valid in an n-dimensional Euclidean vector space $\mathbf{R}^n$ $(n \in N)$:

Fact 4.1.1 *Let μ be a probability Borel measure on $\mathbf{R}$ with a strictly positive continuous distribution function and λ_n be a Lebesgue measure defined on the n-dimensional Euclidean vector space $\mathbf{R}^n$. Then the measures μ^n and λ_n are equivalent.*

Fact 4.1.2 *Let $(\mu_k)_{1 \le k \le n}$ be a family of Borel probability measures on $\mathbf{R}$ with strictly positive continuous distribution functions and λ_n be a Lebesgue measure defined on the n-dimensional Euclidean vector space $\mathbf{R}^n$. Then the measures $\prod_{k=1}^n \mu_k$ and λ_n are equivalent.*

Fact 4.1.3 *Let $(\mu_k)_{1 \le k \le n}$ be a family of different Borel probability measures on $\mathbf{R}$ with strictly positive continuous distribution functions. Then the measures μ_k^n and μ_l^n are equivalent for each $1 \le k \le l \le n$.*

The proof of the above mentioned facts employs the following simple lemma which is well known in the literature.

Lemma 4.1.1 *Let μ_k and ν_k be equivalent non-trivial σ-finite Borel measures on the measurable space (X_k, M_k) for $1 \le k \le n$. Then the measures $\prod_{k=1}^n \mu_k$ and $\prod_{k=1}^n \nu_k$ are equivalent.*

G. Pantsulaia, *Applications of Measure Theory to Statistics*,
DOI 10.1007/978-3-319-45578-5_4

Proof Note that for proving Lemma 4.1.1, it suffices to prove that if μ_k is absolutely continuous with respect to $\nu_k (k = 1, 2)$, then so is $\prod_{k=1}^{2} \mu_k$ with respect to $\prod_{k=1}^{2} \nu_k$.

Assume that $E \in M_1 \times M_2$ such that $\mu_1 \times \mu_2(E) = 0$. We have to show that $\nu_1 \times \nu_2(E) = 0$.

By the Fubini theorem we have

$$0 = \mu_1 \times \mu_2(E) = \int_{X_1} \mu_2(E \cap (\{x\} \times X_2))d\mu_1(x).$$

This means that

$$\mu_1(\{x : \mu_2(E \cap (\{x\} \times X_2)) > 0\}) = 0 \tag{4.1.2}$$

or, equivalently,

$$\mu_1(X_1 \setminus \{x : \mu_2(E \cap (\{x\} \times X_2)) = 0\}) = 0. \tag{4.1.3}$$

Since $\nu_1 \ll \mu_1$, we have

$$\{x : \mu_2(E \cap (\{x\} \times X_2)) = 0\} \subseteq \{x : \nu_2(E \cap (\{x\} \times X_2)) = 0\}. \tag{4.1.4}$$

Since $\nu_1 \ll \mu_1$ and

$$\mu_1(X_1 \setminus \{x : \nu_2(E \cap (\{x\} \times X_2)) = 0\}) = 0, \tag{4.1.5}$$

we have

$$\nu_1(X_1 \setminus \{x : \nu_2(E \cap (\{x\} \times X_2)) = 0\}) = \nu_1(\{x : \nu_2(E \cap (\{x\} \times X_2)) > 0\}) = 0. \tag{4.1.6}$$

Finally, we get

$$\nu_1 \times \nu_2(E) = \int_{X_1} \nu_2(E \cap (\{x\} \times X_2))d\nu_1(x) =$$

$$\int_{\{x:\nu_2(E\cap(\{x\}\times X_2))>0\}} \nu_2(E \cap (\{x\} \times X_2))d\nu_1(x)+$$

$$\int_{\{x:\nu_2(E\cap(\{x\}\times X_2))=0\}} \nu_2(E \cap (\{x\} \times X_2))d\nu_1(x) = 0. \tag{4.1.7}$$

In order to obtain the infinite-dimensional versions of Facts 4.1.1 and 4.1.2, we must know what measures in infinite-dimensional topological vector spaces can be taken as partial analogs of the Lebesgue measure in $\mathbf{R}^n$ $(n \in N)$. In this direction the results of I. Girsanov and B. Mityagyn [GM] and Sudakov [Sud] on the nonexistence of nontrivial translation-invariant σ-finite Borel measures in infinite-dimensional topological vector spaces are important. These authors assert that the properties of σ-finiteness and of translation-invariance are not consistent. Hence one can weaken the property of translation-invariance for analogs of the Lebesgue measure and construct nontrivial σ-finite Borel measures which are invariant under everywhere dense linear manifolds. We wish to make a special note that Moore [M], Yamasaki [Y] and Kharazishvili [Kh] had described the constructions of such measures in an infinite-dimensional Polish topological vector space R^N of all real-valued sequences equipped with product topology, which are invariant under the group $R^{(N)}$ of all eventually zero real-valued sequences. Such measures can be called Moore–Yamasaki–Kharazishvili type measures in R^N. Using Kharazishvili's approach [Kh], it is proved in [GKPP] that every infinite-dimensional Polish linear space admits a σ-finite non-trivial Borel measure that is translation invariant with respect to a dense linear subspace. This extends a recent result of Gill, Pantsulaia and Zachary [GPZ] on the existence of such measures in Banach spaces with Schauder bases.

In this chapter, we focus on the question whether Facts 4.1.1 and 4.1.2 admit infinite-dimensional generalizations in terms of Moore–Yamasaki–Kharazishvili type measures in $\mathbb{R}^{\mathbb{N}}$. To this end, our consideration will involve the following problems.

Problem 4.1.1 Let μ be a probability Borel measure on R with a strictly positive continuous distribution function and λ be a Moore–Yamasaki–Kharazishvili type measure in $\mathbb{R}^{\mathbb{N}}$. Are the measures μ^N and λ equivalent ?

Problem 4.1.2 Let $(\mu_k)_{k\in N}$ be a family of Borel probability measures on R with strictly positive continuous distribution functions and λ be a Moore–Yamasaki–Kharazishvili type measure in $\mathbb{R}^{\mathbb{N}}$. Are the measures $\prod_{k\in N}\mu_k$ and λ equivalent?

Concerning with Fact 4.1.3, it is natural to consider the following problems.

Problem 4.1.3 Let μ_1 and μ_2 be different Borel probability measures on R with strictly positive continuous distribution functions. Are the measures μ_1^N and μ_2^N equivalent?

Problem 4.1.4 Let $(\mu_i)_{i\in I}$ be a family of all Borel probability measures on R with strictly positive continuous distribution functions. Setting $S(R^N) := \cap_{i\in I}\text{dom}(\overline{\mu_i^N})$, where $\overline{\mu_i^N}$ denotes a usual completion of the measure $\mu_i^N (i \in I)$, does there exist a partition $(D_i)_{i\in I}$ of R^N into elements of the σ-algebra $S(R^N)$ such that $\overline{\mu_i^N}(D_i) = 1$ for each $i \in I$?

Problems 4.1.3 and 4.1.4 are not new and have been investigated by many authors in more general formulations. In this direction, we should specially mention the result of S. Kakutani [Kak] (see Theorem 4.3) stating that if one has equivalent probability measures μ_i and ν_i on the σ-algebra $\mathbf{L}_i$ of subsets of a set Ω_i, $i = 1, 2, \dots$ and if μ and ν denote respectively the infinite product measures $\prod_{i\in N}\mu_i$ and $\prod_{i\in N}\nu_i$ on the infinite product σ-algebra generated on the infinite product set Ω, then μ and ν are either equivalent or orthogonal. Similar dichotomies have revealed themselves in the study of Gaussian stochastic processes. C. Cameron and W.E. Martin [CM] proved that if one considers the measures induced on a path space by a Wiener process on the unit interval, then, if the variances of corresponding processes are different, the measures are orthogonal. Results of this kind were generalized by many authors (cf. [F, G] and others). A.M. Vershik [V] proved that a group of all admissible translations (in the sense of quasiinvariance) of an arbitrary Gaussian measure, defined in an infinite-dimensional separable Hilbert space, is a linear manifold.

For study of the general problem of equivalence and singularity of two product measures was carried out by various authors using different approaches, among which are the strong law of large numbers, the properties of the Hellinger integral [H], the zero-one laws [LM] and so on. In this chapter, we propose a new approach for the solution of Problems 4.1.3 and 4.1.4, which uses the properties of uniformly distributed sequences [KN].

In Sects. 4.2 and 4.3, we give solutions of Problems 4.1.1 and 4.1.2 which are due to Yamasaki [Y]. In Sect. 4.4, we give solutions of Problems 4.1.3 and 4.1.4.

4.2 Solution of the Problem 4.1.1

A negative solution of Problem 4.1.1 is contained in the following

Fact 4.2.1 ([Y], Proposition 2.1, p. 696) *Let $f(x)$ be a measurable function on R^1 which satisfies $f(x) > 0$ and $\int_{-\infty}^{+\infty} f(x)dx = 1$. Let μ be the stationary product measure of f (i.e. $d\mu = \prod_{i=1}^{\infty} f(x_i)dx_i$) and $\mathbf{R}^{(\mathbf{N})}$ be a linear vector space of all eventually zero real-valued sequences. Then μ is $\mathbf{R}^{(\mathbf{N})}$-quasi-invariant but μ has no equivalent Moore–Yamasaki–Kharazishvili type measure.*

Proof Let $\sum$ be the permutation group on the set of all natural numbers $N = \{1, 2, \dots\}$. $\sum$ can be regarded as a transformation group on $\mathbf{R}^{\mathbf{N}}$, and μ is $\sum$-invariant. Let $\sum_0$ be the subgroup of $\sum$ generated by all transpositions (of two elements of N). $\sum_0$ consists of such a permutation $\sigma \in \sum$ that satisfies $\sigma(i) = i$ except finite numbers of $i \in N$. As shown in [Sh], the measure μ is $\sum_0$-ergodic.

Now, we shall derive a contradiction assuming that μ has an equivalent $\mathbf{R}^{(\mathbf{N})}$-invariant σ-finite measure ν. Since $\mu \approx \nu$, where μ is $\sum_0$-invariant and $\sum_0$-ergodic, and ν is $\sum_0$-invariant, we have that $\mu = c\nu$ for some constant $c > 0$. Thus, the $\mathbf{R}^{(\mathbf{N})}$-invariance of ν implies that of μ, which is a contradiction.

Therefore it suffices to prove that ν is $\sum_0$-invariant, namely for each $\sigma \in \sum_0$, $\tau_\sigma \nu = \nu$, where

$$\tau_\sigma \nu(B) = \nu(\sigma^{-1}(B)), \tag{4.2.1}$$

for each $B \in \mathbb{B}(R^N)$. Since $\tau_\sigma \mu = \mu$, we have $\tau_\sigma \nu \approx \nu$. On the other hand, ν is $\mathbf{R}^{(\mathbf{N})}$-ergodic because μ is such. Therefore if $\tau_\sigma \nu$ is $\mathbf{R}^{(\mathbf{N})}$-invariant, then we have $\tau_\sigma \nu = c_\sigma \nu$ for some constant $c_\sigma > 0$. In particular for a transposition σ, $\sigma^2 = l$ implies $c_\sigma^2 = l$, hence $c_\sigma = 1$. This means that ν is invariant under any transposition. Since $\sum_0$ is generated by the set of all transpositions, we have proved the $\sum_0$-invariance of ν.

To complete the proof of Fact 4.2.1, it remains only to prove that $\tau_\sigma \nu$ is $\mathbf{R}^{(\mathrm{N})}$-invariant. Since ν is $\mathbf{R}^{(\mathrm{N})}$ -invariant, we have $\tau_x \nu = \nu$ for any $x \in \mathbf{R}^{(\mathrm{N})}$, where $\tau_x \nu(B) = \nu(B - x)$ for each $B \in \mathscr{R}^{\mathscr{N}}$. Therefore

$$(\forall x)(x \in \mathbf{R}^{(\mathrm{N})} \to \tau_\sigma \tau_x \nu = \tau_\sigma \nu). \tag{4.2.2}$$

However, we can easily show $\tau_\sigma \tau_x \nu = \tau_{\sigma x} \tau_\sigma \nu$, so (4.2.2) implies that $\tau_\sigma \nu$ is $\sigma(\mathbf{R}^{(\mathrm{N})})$-invariant. Since σ maps $\mathbf{R}^{(\mathrm{N})}$ onto $\mathbf{R}^{(\mathrm{N})}$, namely $\sigma(\mathbf{R}^{(\mathrm{N})}) = \mathbf{R}^{(\mathrm{N})}$ we have proved the $\mathbf{R}^{(\mathrm{N})}$-invariance of $\tau_\sigma \nu$.

4.3 Particular Solution of Problem 4.1.2

Remark 4.3.1 If in the formulation of Problem 4.1.2 the equality $\mu_k = \mu_n$ holds true for each $k, n \in N$, then Problem 4.1.2 is reduced to Problem 4.1.1. Following Fact 4.1.2 we know that the answer to Problem 4.1.1 is *no*.

In order to answer positively to Problem 4.1.2, we give the construction of Moore–Yamasaki–Kharazishvili measures and formulate their some properties.

Example 4.3.1 ([P], *Section 1, p. 354*) Let $\mathbf{R}^N$ be the topological vector space of all real-valued sequences equipped with the Tychonoff topology. Let us denote by $B(\mathbf{R}^N)$ the σ-algebra of all Borel subsets in $\mathbf{R}^N$.

Let $(a_i)_{i \in N}$ and $(b_i)_{i \in N}$ be sequences of real numbers such that

$$(\forall i)(i \in N \to a_i < b_i). \tag{4.3.1}$$

We put

$$A_n = \mathbf{R}_0 \times \cdots \times \mathbf{R}_n \times \left(\prod_{i>n} \Delta_i \right), \tag{4.3.2}$$

for $n \in N$, where

$$(\forall i)(i \in N \to \mathbf{R}_i = \mathbf{R} \;\&\; \Delta_i = [a_i; b_i[). \tag{4.3.3}$$

We put also

$$\Delta = \prod_{i \in N} \Delta_i. \tag{4.3.4}$$

For an arbitrary natural number $i \in N$, consider the Lebesgue measure μ_i defined on the space $\mathbf{R}_i$ and satisfying the condition $\mu_i(\Delta_i) = 1$. Let us denote by λ_i the normalized Lebesgue measure defined on the interval Δ_i.

For an arbitrary $n \in N$, let us denote by ν_n the measure defined by

$$\nu_n = \prod_{1 \le i \le n} \mu_i \times \prod_{i > n} \lambda_i, \tag{4.3.5}$$

and by $\overline{\nu}_n$ the Borel measure in the space $\mathbf{R}^N$ defined by

$$(\forall X)(X \in B(\mathbf{R}^N) \to \overline{\nu}_n(X) = \nu_n(X \cap A_n)). \tag{4.3.6}$$

Note that (see [P], Lemma 1.1, p. 354) for an arbitrary Borel set $X \subseteq \mathbf{R}^N$ there exists a limit

$$\nu_\Delta(X) = \lim_{n \to \infty} \overline{\nu}_n(X). \tag{4.3.7}$$

Moreover, the functional ν_Δ is a nontrivial σ-finite measure defined on the Borel σ-algebra $B(\mathbf{R}^N)$.

Recall that an element $h \in \mathbf{R}^N$ is called an admissible translation in the sense of invariance for the measure ν_Δ if

$$(\forall X)(X \in B(\mathbf{R}^N) \to \nu_\Delta(X + h) = \nu_\Delta(X)). \tag{4.3.8}$$

We define

$$G_\Delta = \{h : h \in \mathbf{R}^N \ \& \ h \text{ is an admissible translation for } \nu_\Delta\}. \tag{4.3.9}$$

It is easy to show that G_Δ is a vector subspace of $\mathbf{R}^N$.

We have the following

Lemma 4.3.1 ([P], Theorem 1.4, p.356) *The following conditions are equivalent:*

$$(1)\ g = (g_1, g_2, \dots) \in G_\Delta, \tag{4.3.10}$$

$$(2)\ (\exists n_g)(n_g \in N \to \textit{ the series} \sum_{i \ge n_g} \ln(1 - \frac{|g_i|}{b_i - a_i}) \textit{ is convergent}). \tag{4.3.11}$$

Let $\mathbf{R}^{(N)}$ be the space of all eventually zero sequences, i.e.,

$$\mathbf{R}^{(N)} = \{(g_i)_{i \in N} | (g_i)_{i \in N} \in \mathbf{R}^N \ \& \ \text{card}\{i | g_i \neq 0\} < \aleph_0\}. \tag{4.3.12}$$

It is clear that, on the one hand, for an arbitrary compact infinite-dimensional parallelepiped $\Delta = \prod_{k \in N} [a_k, b_k]$, we have

$$\mathbf{R}^{(N)} \subset G_\Delta. \tag{4.3.13}$$

On the other hand, $G_{\Delta} \setminus \mathbf{R}^{(N)} \neq \emptyset$ since an element $(g_i)_{i \in N}$ defined by

$$(\forall i) \left(i \in N \to g_i = \left(1 - \exp\left\{ -\frac{b_i - a_i}{2^i} \right\} \times (b_i - a_i) \right) \right) \tag{4.3.14}$$

belongs to the difference $G_{\Delta} \setminus \mathbf{R}^{(N)}$.

It is easy to show that the vector space G_{Δ} is everywhere dense in $\mathbf{R}^N$ with respect to the Tychonoff topology since $\mathbf{R}^{(N)} \subset G_{\Delta}$.

Below we present an example of the product of an infinite family of Borel probability measures on R with strictly positive continuous distribution functions and a Moore–Yamasaki–Kharazishvili type measure in $\mathbb{R}^{\mathbb{N}}$, such that these measures are equivalent.

Let $(c_n)_{n \in N}$ be a sequence of positive numbers such that $0 < c_n < l$. On the real axis R, for each n consider a continuous function $f_n(x)$ which satisfies:

$$0 < f_n(x) < 1, \int_{-\infty}^{+\infty} f_n(x) dx = 1, \tag{4.3.15}$$

$$f_n(x) = c_k \text{ for } x \in [0, 1]. \tag{4.3.16}$$

Such a function $f_n(x)$ exists certainly for any $n \in N$.

For $n \in N$, let us denote by μ_n a Borel probability measure on R defined by the distribution density function f_n.

Fact 4.3.1 *If $\prod_{n \in N} c_n > 0$, then the measures $\prod_{n \in N} \mu_n$ and $\nu_{[0,1]^N}$ are equivalent.*

Proof By the Fubini theorem, one can easily prove that the measure $\prod_{n \in N} \mu_n$ is $\mathbf{R}^{(N)}$-quasiinvariant. According to [Sh], every product measure on $\mathbf{R}^N$ is $\mathbf{R}^{(N)}$-ergodic. Therefore, $\prod_{n \in N} \mu_n$, hence $\nu_{[0,1]^N}$, too is $\mathbf{R}^{(N)}$- ergodic.

For $x = (x_n) \in \mathbf{R}^N$, define a function $f(x)$ by:

$$f(x) = \prod_{n \in N} f_n(x). \tag{4.3.17}$$

Since $0 < f(x_n) < 1$, the partial product decreases monotonically, so that the infinite product in (4.3.1) exists certainly. If $x \in A_n$, then $x_k \in [0, 1]$ for $k > n$, so we have

$$f(x) = \prod_{k=1}^{n} f_k(x_k) \prod_{k>n} c_k > 0. \tag{4.3.18}$$

Thus $f(x)$ is positive on A_n, hence positive on $\cup_{n \in N} A_n$, too. On the other hand, since $\nu_{[0,1]^N}(\mathbf{R}^N \setminus \cup_{n \in N} A_n) = 0$, we see that $f(x)$ is positive for $\nu_{[0,1]^N}$-almost all x.

Now, define a measure ν' on $\mathbf{R}^N$ by

$$\nu'(X) = \int_X f(x) d\nu_{[0,1]^N}(x) \tag{4.3.19}$$

for $X \in \mathbf{B}(\mathbf{R}^N)$.

Let us show that $\prod_{n \in N} \mu_n = \nu'$. For this it suffices to show that for each $A \in \mathbf{B}(R^n)$ we have

$$\nu'(A \times \mathbf{R}^{N \setminus \{1,\dots,n\}}) = \prod_{n \in N} \mu_n(A \times \mathbf{R}^{N \setminus \{1,\dots,n\}}). \tag{4.3.20}$$

Indeed, we have

$$\begin{aligned}
\nu'(A \times \mathbf{R}^{N \setminus \{1,\dots,n\}}) &= \int_{A \times \mathbf{R}^{N \setminus \{1,\dots,n\}}} f(x) d\nu_{[0,1]^N}(x) \\
&= \lim_{m \to +\infty} \int_{A_m \cap (A \times \mathbf{R}^{N \setminus \{1,\dots,n\}})} f(x) d\nu_{[0,1]^N}(x) \\
&= \lim_{m \to +\infty} \int_{A \times \prod_{k=n+1}^m \mathbf{R} \times \prod_{k>m}[0,1]} f(x) d\nu_{[0,1]^N}(x) \\
&= \lim_{m \to +\infty} \int_{A \times \prod_{k=n+1}^m \mathbf{R} \times \prod_{k>m}[0,1]} f(x) d\left(\prod_{k=1}^m \mu_k \times \prod_{k>m} \lambda_k\right) \\
&= \lim_{m \to +\infty} \int_{A \times \prod_{k=n+1}^m \mathbf{R}} \left(\int_{\prod_{k>m}[0,1]} f(x) d\prod_{k>m} \lambda_k\right) d\prod_{k=1}^m \mu_k \\
&= \lim_{m \to +\infty} \int_{\prod_{k>m}[0,1]} \prod_{k>m} f_k(x_k) d\prod_{k>m} \lambda_k \\
&\times \lim_{m \to +\infty} \int_{A \times \prod_{k=n+1}^m \mathbf{R}} \prod_{k=1}^m f_k(x_k) d\prod_{k=1}^m \mu_k \\
&= \lim_{m \to +\infty} \int_{\prod_{k>m}[0,1]} \prod_{k>m} f_k(x_k) d\prod_{k>m} \lambda_k \\
&\times \lim_{m \to +\infty} \int_A \prod_{k=1}^n f_k(x_k) d\prod_{k=1}^n \mu_k \times \int_{\prod_{k=n+1}^m \mathbf{R}} \prod_{k=n+1}^m f_k(x_k) d\prod_{k=n+1}^m \mu_k \\
&= \lim_{m \to +\infty} \prod_{k>m} c_k \times \prod_{k=1}^n \mu_k(A) = \prod_{k=1}^n \mu_k(A) = \prod_{k \in N} \mu_k(A \times \mathbf{R}^{N \setminus \{1,\dots,n\}}).
\end{aligned} \tag{4.3.21}$$

This ends the proof of Fact 4.3.1.

Remark 4.3.2 Let the product-measure $\prod_{k \in N} \mu_k$ comes from Fact 4.3.1. Then by virtue of Lemma 4.3.1, we know that the group of all admissible translations (in the sense of invariance) for the measure $\nu_{[0,1]^N}$ is $l_1 = \{(x_k)_{k \in N} : (x_k)_{k \in N} \in$

R^N & $\sum_{k \in N} |x_k| < +\infty\}$. Following Fact 4.3.1, the measures $\prod_{k \in N} \mu_k$ and $\nu_{[0,1]^N}$ are equivalent, which implies that the group of all admissible translations (in the sense of quasiinvariance) for the measure $\prod_{k \in N} \mu_k$ is equal to l_1.

For $(x_k)_{k \in N} \in l_1$, we set $\nu_k(X) = \mu_k(X - x_k)$ for each $X \in B(R)$. It is obvious that μ_k and ν_k are equivalent for each $k \in N$. For $k \in N$ and $x \in R$, we put $\rho_k(x) = \frac{d\nu_k(x)}{d\mu_k(x)}$. Let us consider the product-measures $\mu = \prod_{k \in N} \mu_k$ and $\nu = \prod_{k \in N} \nu_k$. On the one hand, following our observation, the measures μ and ν are equivalent. On the other hand, by virtue of Kakutani's well known result (see, [Kak]), since the measures μ and ν are equivalent, we deduce that the infinite product $\prod_{k \in N} \alpha_k$ is divergent to zero, where $\alpha_k = \int_R \sqrt{\rho_k(x_k)} d\mu_k(x_k)$. In this case $r_n(x) = \prod_{k=1}^{n} \rho_k(x)$ is convergent (in the mean) to the function $r(x) = \prod_{k=1}^{\infty} \rho_k(x)$ which is the density of the measure ν with respect to μ, i.e.,

$$r(x) = \frac{d\nu(x)}{d\mu(x)}. \tag{4.3.22}$$

Remark 4.3.3 The approach used in the proof of Fact 4.3.1 is taken from [Y] (see Proposition 4.1, p. 702).

In the context of Fact 4.3.1 we state the following

Problem 4.3.1 Do there exist a family $(\mu_k)_{k \in N}$ of linear Gaussian probability measures on R and a Moore–Yamasaki–Kharazishvili type measure λ in $\mathbf{R^N}$ such that the measures $\prod_{k \in N} \mu_k$ and λ are equivalent?

4.4 Solution of Problems 4.1.3 and 4.1.4

We present a new approach for the solution of Problems 4.1.3 and 4.1.4, which is quite different from the approach introduced in [Kak]. Our approach uses the technique of the so-called uniformly distributed sequences. The main notions and auxiliary propositions are taken from [KN].

Definition 4.4.1 ([KN]) A sequence $(x_k)_{k \in N}$ of real numbers from the interval (a, b) is said to be equidistributed or uniformly distributed on an interval (a, b) if for any subinterval $[c, d]$ of (a, b) we have

$$\lim_{n \to \infty} n^{-1} \#(\{x_1, x_2, \dots, x_n\} \cap [c, d]) = (b - a)^{-1}(d - c), \tag{4.4.1}$$

where # denotes the counting measure.

Now let X be a compact Polish space and μ be a probability Borel measure on X. Let $\mathbf{R}(X)$ be a space of all bounded continuous measurable functions defined on X.

Definition 4.4.2 A sequence $(x_k)_{k \in N}$ of elements of X is said to be μ-equidistributed or μ-uniformly distributed on X if for every $f \in \mathbf{R}(X)$ we have

$$\lim_{n\to\infty} n^{-1} \sum_{k=1}^{n} f(x_k) = \int_X f d\mu. \tag{4.4.2}$$

Lemma 4.4.1 ([KN], Lemma 2.1, p. 199) *Let $f \in \mathbf{R}(X)$. Then, for μ^N-almost every sequence $(x_k)_{k \in N} \in X^N$, we have*

$$\lim_{n\to\infty} n^{-1} \sum_{k=1}^{n} f(x_k) = \int_X f d\mu. \tag{4.4.3}$$

Lemma 4.4.2 ([KN], pp. 199–201) *Let S be a set of all μ-equidistributed sequences on X. Then we have $\mu^N(S) = 1$.*

Corollary 4.4.1 ([ZPS], Corollary 2.3, p. 473) *Let ℓ_1 be a Lebesgue measure on $(0, 1)$. Let D be a set of all ℓ_1-equidistributed sequences on $(0, 1)$. Then we have $\ell_1^N(D) = 1$.*

Definition 4.4.3 Let μ be a probability Borel measure on R with a distribution function F. A sequence $(x_k)_{k \in N}$ of elements of R is said to be μ-equidistributed or μ-uniformly distributed on R if for every interval $[a, b](-\infty \leq a < b \leq +\infty)$ we have

$$\lim_{n\to\infty} n^{-1}\#([a, b] \cap \{x_1, \dots, x_n\}) = F(b) - F(a). \tag{4.4.4}$$

Lemma 4.4.3 ([ZPS], Lemma 2.4, p. 473) *Let $(x_k)_{k \in N}$ be an ℓ_1-equidistributed sequence on $(0, 1)$, F be a strictly increasing continuous distribution function on R and p be a Borel probability measure on R defined by F. Then $(F^{-1}(x_k))_{k \in N}$ is p-equidistributed on R.*

Corollary 4.4.2 ([ZPS], Corollary 2.4, p. 473) *Let F be a strictly increasing continuous distribution function on R and p be a Borel probability measure on R defined by F. Then for a set $D_F \subset R^N$ of all p-equidistributed sequences on R we have:*
(i) $D_F = \{(F^{-1}(x_k))_{k \in N} : (x_k)_{k \in N} \in D\}$;
(ii) $p^N(D_F) = 1$.

Lemma 4.4.4 *Let F_1 and F_2 be different strictly increasing continuous distribution functions on R, and p_1 and p_2 be Borel probability measures on R defined by F_1 and F_2, respectively. Then there does not exist a sequence of real numbers $(x_k)_{k \in \mathbf{N}}$ which simultaneously is p_1-equidistributed and p_2-equidistributed.*

Proof Assume the contrary and let $(x_k)_{k \in \mathbf{N}}$ be such a sequence. Since F_1 and F_2 are different, there is a point $x_0 \in \mathbb{R}$ such that $F_1(x_0) \neq F_2(x_0)$. The latter relation is not possible under our assumption because $(x_k)_{k\in \mathbf{N}}$ simultaneously is p_1-equidistributed and p_2-equidistributed, which implies

$$F_1(x_0) = \lim_{n\to\infty} n^{-1}\#((-\infty, x_0] \cap \{x_1, \ldots, x_n\}) = F_2(x_0). \tag{4.4.5}$$

The next theorem contains the solution of Problem 4.1.3.

Theorem 4.4.1 *Let F_1 and F_2 be different strictly increasing continuous distribution functions on R and p_1 and p_2 be Borel probability measures on R, defined by F_1 and F_2, respectively. Then the measures p_1^N and p_2^N are orthogonal.*

Proof Let D_{F_1} and D_{F_2} denote p_1-equidistributed and p_2-equidistributed sequences on R, respectively. By Lemma 4.4.4 we know that $D_{F_1} \cap D_{F_2} = \emptyset$. By Lemma 4.4.2 we know that $p_1^N(D_{F_1}) = 1$ and $p_2^N(D_{F_2}) = 1$. This ends the proof of the theorem.

Definition 4.4.4 Let $\{\mu_i : i \in I\}$ be a family of probability measures defined on a measure space (X, M). Let $S(X)$ be defined by

$$S(X) = \cap_{i\in I}\mathrm{dom}(\overline{\mu}_i),$$

where $\overline{\mu}_i$ denotes a usual completion of the measure μ_i. We say that the family $\{\mu_i : i \in I\}$ is strongly separated if there exists a partition $\{C_i : i \in I\}$ of the space X into elements of the σ-algebra $S(X)$ such that $\overline{\mu}_i(C_i) = 1$ for each $i \in I$.

Definition 4.4.5 Let $\{\mu_i : i \in I\}$ be a family of probability measures defined on a measure space (X, M). Let $S(I)$ denote a minimal σ-algebra generated by singletons of I and the σ-algebra $S(X)$ of subsets of X be defined by

$$S(X) = \cap_{i\in I}\mathrm{dom}(\overline{\mu}_i),$$

where $\overline{\mu}_i$ denotes a usual completion of the measure μ_i for $i \in I$. We say that a $(S(X), S(I))$-measurable mapping $T : X \to I$ is a well-founded estimate of an unknown parameter i $(i \in I)$ for the family $\{\mu_i : i \in I\}$ if the following condition

$$(\forall i)(i \in I \to \mu_i(T^{-1}(\{i\}) = 1) \tag{4.4.6}$$

holds true.

One can easily get the validity of the following assertion.

Lemma 4.4.5 ([ZPS], Lemma 2.5, p. 474) *Let $\{\mu_i : i \in I\}$ be a family of probability measures defined on a measure space (X, M). The following propositions are equivalent:*

(i) The family of probability measures $\{\mu_i : i \in I\}$ is strongly separated;
(ii) There exists a well-founded estimate of an unknown parameter i $(i \in I)$ for the family $\{\mu_i : i \in I\}$.

The next theorem contains the solution of Problem 4.1.4.

Theorem 4.4.2 *Let* $\mathbf{F}$ *be a family of all strictly increasing and continuous distribution functions on* $\mathbf{R}$ *and* p_F *be a Borel probability measure on* R *defined by* F *for each* $F \in \mathbf{F}$. *Then the family of Borel probability measures* $\{p_F^N : F \in \mathbf{F})\}$ *is strongly separated.*

Proof We denote by D_F the set of all p_F-equidistributed sequences on R for each $F \in \mathbf{F}$. By Lemma 4.4.4 we know that $D_{F_1} \cap D_{F_2} = \emptyset$ for each different $F_1, F_2 \in \mathbf{F}$. By Lemma 4.4.2 we know that $p_F^N(D_F) = 1$ for each $F \in \mathbf{F}$. Let us fix $F_0 \in \mathbf{F}$ and define a family $(C_F)_{F\in\mathbf{F}}$ of subsets of $\mathbf{R}^\mathbf{N}$ as follows: $C_F = D_F$ for $F \in \mathbf{F}\backslash\{F_0\}$ and $C_{F_0} = R^N \backslash \cup_{F\in\mathbf{F}\backslash\{F_0\}} D_F$. Since D_F is a Borel subset of $\mathbf{R}^\mathbf{N}$ for each $F \in \mathbf{F}$, we claim that $C_F \in S(\mathbf{R}^\mathbf{N})$ for each $F \in \mathbf{F}\backslash\{F_0\}$. Since $\overline{p_F^N}(R^N \backslash \cup_{F\in\mathbf{F}} D_F) = 0$ for each $F \in \mathbf{F}$, we deduce that $R^N \backslash \cup_{F\in\mathbf{F}} D_F \in \cap_{F\in\mathbf{F}}\mathrm{dom}(\overline{p_F^N}) = S(R^N)$. Since $S(\mathbf{R}^\mathbf{N})$ is an σ-algebra, we claim that $C_{F_0} \in S(R^N)$ because $\overline{p_F^N}(R^N \backslash \cup_{F\in\mathbf{F}} D_F) = 0$ for each $F \in \mathbf{F}$(equivalently, $R^N \backslash \cup_{F\in\mathbf{F}} D_F \in S(R^N)$), $D_{F_0} \in S(R^N)$ and

$$C_{F_0} = R^N \backslash \cup_{F\in\mathbf{F}\backslash\{F_0\}} D_F = (R^N \backslash \cup_{F\in\mathbf{F}} D_F) \cup D_{F_0}. \tag{4.4.7}$$

This ends the proof of the theorem.

By virtue of the results of Lemma 4.4.5 and Theorem 4.4.2 we get the following

Corollary 4.4.3 *Let* $\mathbf{F}$ *be a family of all strictly increasing and continuous distribution functions on* $\mathbf{R}$. *Then there exists a well-founded estimate of an unknown distribution function* F $(F \in \mathbf{F})$ *for the family of Borel probability measures* $\{p_F^N : F \in \mathbf{F}\}$.

Remark 4.4.1 The validity of Theorem 4.4.2 and Corollary 4.4.3 can be obtained for an arbitrary family of strictly increasing and continuous distribution functions on $\mathbf{R}$. Note that Corollary 4.4.3 extends the main result established in [ZPS] (see Lemma 2.6, p. 476).

Remark 4.4.2 The requirements in Theorem 4.4.2 that all Borel probability measures on $\mathbf{R}$ are defined by strictly increasing and continuous distribution functions on $\mathbf{R}$ and the measures under consideration are infinite powers of the corresponding measures are essential. Indeed, let μ be a linear Gaussian measure on $\mathbf{R}$ whose density distribution function has the form $f(x) = \frac{1}{\sqrt{2\pi}}e^{-\frac{x^2}{2}}$ $(x \in \mathbf{R})$. Let δ_x be a Dirac measure defined on the Borel σ-algebra of subsets of $\mathbf{R}$ and concentrated at x $(x \in \mathbf{R})$. Let D be a subset of $\mathbf{R}^\mathbf{N}$ defined by

$$D = \left\{(x_k)_{k\in N} : \lim_{n\to\infty} \frac{\sum_{k=1}^n x_k}{n} = 0\right\}. \tag{4.4.8}$$

It is obvious that D is a Borel subset of $\mathbf{R}^\mathbf{N}$. For $(x_k)_{k\in N} \in D$ we set $\mu_{(x_k)_{k\in N}} = \prod_{k\in N} \delta_{x_k}$.[1] Let us consider the family of Borel probability measures $\{\mu^N\} \cup \{\mu_{(x_k)_{k\in N}} :$

[1]Note that $\prod_{k\in N} \delta_{x_k} = \delta_{(x_k)_{k\in N}}$.

$(x_k)_{k\in N} \in D\}$. It is obvious that it is an orthogonal family of Borel product-measures which is not strong separable. Indeed, assume the contrary and let $\{C\} \cup \{C_{(x_k)_{k\in N}} : (x_k)_{k\in N} \in D\}$ be such a partition of $\mathbf{R}^{\mathbf{N}}$ into elements of the σ-algebra $S_0(R^N) = \cap_{(x_k)_{k\in N}\in D}\mathrm{dom}(\overline{\mu}_{(x_k)_{k\in N}}) \cap \mathrm{dom}(\overline{\mu^N})$ that $\overline{\mu}_{(x_k)_{k\in N}}(C_{(x_k)_{k\in N}}) = 1$ for $(x_k)_{k\in N} \in D$ and $\overline{\mu^N}(C) = 1$. Since $(x_k)_{k\in N} \in C_{(x_k)_{k\in N}}$ for each $(x_k)_{k\in N} \in D$ we deduce that $D \cap C = \emptyset$. This implies that $\overline{\mu^N}(C) \le \overline{\mu^N}(\mathbf{R}^{\mathbf{N}} \setminus D) = 0$ because by the strong law of large numbers we have that $\overline{\mu^N}(D) = 1$. The latter relation is a contradiction and Remark 4.4.2 is proved.

Remark 4.4.3 By using Glivenko–Canteli theorem we can obtain the solution of Problem 1.4 in more general formulation. More precisely, if $\mathbf{F}$ is any family of different distribution functions on $\mathbf{R}$ and p_F denotes Borel probability measure on R defined by F for each $F \in \mathbf{F}$, then the family of Borel probability measures $\{p_F^\infty : F \in \mathbf{F})\}$ is strongly separated. Indeed, for $F \in \mathbf{F}$ we put

$$D_F = \left\{(x_k)_{k\in N} : (x_k)_{k\in N} \in R^\infty \ \& \ \lim_{n\to\infty} \sup_{x\in R} \left|\frac{\#(\{x_1, \dots, x_n\} \cap (-\infty, x])}{n} - F(x)\right| = 0\right\}.$$

By Glivenko–Canteli theorem we get

$$P_F^\infty(D_F) = 1.$$

Now let show that $D_{F_1} \cap D_{F_2} = \emptyset$ for different $F_1, F_2 \in \mathbf{F}$. Indeed, assume the contrary and let $(x_k)_{k\in N} \in D_{F_1} \cap D_{F_2}$. Let $x_0 \in \mathbf{R}$ be such a point that $F_1(x_0) \neq F_2(x_0)$. Then for each $n \in N$ we get

$$\begin{aligned}
|F_2(x_0) - F_1(x_0)| &= \left|\left(F_2(x_0) - \frac{\#(\{x_1, \dots, x_n\} \cap (-\infty, x_0])}{n}\right)\right. \\
&\quad \left. - \left(F_1(x_0) - \frac{\#(\{x_1, \cdots, x_n\} \cap (-\infty, x_0])}{n}\right)\right| \\
&\le \left|F_2(x_0) - \frac{\#(\{x_1, \dots, x_n\} \cap (-\infty, x_0])}{n}\right| \\
&\quad + \left|F_1(x_0) - \frac{\#(\{x_1, \dots, x_n\} \cap (-\infty, x_0])}{n}\right| \\
&\le \sup_{x\in\mathbf{R}} \left|F_2(x) - \frac{\#(\{x_1, \dots, x_n\} \cap (-\infty, x])}{n}\right| \\
&\quad + \sup_{x\in\mathbf{R}} \left|F_1(x) - \frac{\#(\{x_1, \dots, x_n\} \cap (-\infty, x])}{n}\right|. \qquad (4.4.9)
\end{aligned}$$

Finally we get

$$
\begin{aligned}
&|F_2(x_0) - F_1(x_0)| \\
&\leq \lim_{n\to\infty} \sup_{x\in\mathbf{R}} |F_2(x) - \frac{\#(\{x_1,\ldots,x_n\}\cap(-\infty,x])}{n}| \\
&+ \lim_{n\to\infty} \sup_{x\in\mathbf{R}} |F_1(x) - \frac{\#(\{x_1,\ldots,x_n\}\cap(-\infty,x])}{n}| = 0,
\end{aligned} \tag{4.4.10}
$$

which is the contradiction. Remark 4.4.3 will be proved if we will use the construction used in the proof of Theorem 4.4.2.

References

[CM] Cameron, R.H., Martin, W.T.: On Transformations of Wiener integrals under translations. Ann. Math **45**, 386–396 (1944)

[F] Feldman, J.: Equivalence and orthogonality of Gaussian processes. Pacific J. Math. **8**, 699–708 (1958)

[GKPP] Gill, T., Kirtadze, A., Pantsulaia, G., Plichko, A.: Existence and uniqueness of translation invariant measures in separable Banach spaces. Funct. Approx. Comment. Math. **50**(2), 401–419 (2014)

[GPZ] Gill, T.L., Pantsulaia, G.R., Zachary, W.W.: Constructive analysis in infinitely many variables. Commun. Math. Anal. **13**(1), 107–141 (2012)

[GM] Girsanov, I.V., Mityasin, B.S.: Quasi-invariant measures and linear topological spaces (in Russian). Nauchn. Dokl. Vys. Skol. **2**, 5–10 (1959)

[G] Grenander, U.: Stochastic processes and statistical inference. Ark. Mat. **1**, 195–277 (1950)

[H] Hill, D.G.B.: σ-finite invariant measures on infinite product spaces. Trans. Amer. Math. Soc. **153**, 347–370 (1971)

[Kak] Kakutani, S.: On equivalence of infinite product measures. Ann. Math. **4**(9), 214–224 (1948)

[Kh] Kharazishvili, A.B.: On invariant measures in the Hilbert space (in Russian). Bull. Acad. Sci. Georgian SSR. **114**(1), 41–48 (1984)

[KKP] Kintsurashvili, M., Kiria, T., Pantsulaia, G.: On Moore-Yamasaki-Kharazishvili type measures and the infinite powers of Borel diffused probability measures on **R**. http://arxiv.org/pdf/1506.02935v1.pdf

[KN] Kuipers, L., Niederreiter, H.: Uniform Distribution of Sequences. Wiley, New York (1974)

[LM] LePage, R.D., Mandrekar, V.: Equivalence-singularity dichotomies from zero-one laws. Proc. Am. Math. Soc. **31**, 251–254 (1972)

[M] Moore, C.C.: Invariant measures on product spaces. In: Proceedings of the Fifth Berkeley Symposium Mathematical Statistics and Probability. Vol. II: Contributions to Probability Theory Part, vol. 2, pp. 447–459 (1965–1966)

[P] Pantsulaia, G.: Duality of measure and category in infinite-dimensional separable Hilbert space l_2. Int. J. Math. Math. Sci. **30**(6), 353–363 (2002)

[Sh] Shimomura, H.: An aspect of quasi-invariant measures on R^∞. Publ. Res. Inst. Math. Sci. **11**(3), 749–773 (1975/1976)

[Sk] Skorokhod, A.V.: Integration in Hilbert space (in Russian), Moscow (1974). (Springer, English transl. (1975))

[Sud] Sudakov, V.N.: Linear sets with quasi-invariant measure (in Russian). Dokl. Akad. Nauk SSSR. **127**, 524–525 (1959)

[V] Veršik, A.M.: Duality in the theory of measure in linear spaces (in Russian). Dokl. Akad. Nauk SSSR. **170**, 497–500 (1966)

[X] Xia, D.X.: Measure and integration on infinite-dimensional spaces. Academic Press, New York (1972)

[Y] Yamasaki, Y.: Translationally invariant measure on the infinite-dimensional vector space. Publ. Res. Inst. Math. Sci. **16**(3), 693–720 (1980)

[ZPS] Zerakidze, Z., Pantsulaia, G., Saatashvili, G.: On the separation problem for a family of Borel and Baire G-powers of shift-measures on $\mathbb{R}$. Ukr. Math. J. **65**(4), 470–485 (2013)

Chapter 5
Objective and Strong Objective Consistent Estimates of Unknown Parameters for Statistical Structures in a Polish Group Admitting an Invariant Metric

5.1 Introduction

In order to explain the big gap between the theory of mathematical statistics and results of hypothesis testing, concepts of subjective and objective infinite sample consistent estimates of a useful signal in the linear one-dimensional stochastic model were introduced in [PK1]. This approach essentially used the concept of Haar null sets in Polish topological vector spaces introduced by J.P.R. Christensen [Chr1].

The Polish topological vector space $\mathbf{R}^N$ of all real-valued sequences (equivalently, of infinite samples) equipped with the Tychonoff metric plays a central role in the theory of statistical decisions because a definition of any consistent estimate of an unknown parameter in various stochastic models without infinite samples is simply impossible.

From the point of view of the theory of Haar null sets in $\mathbf{R}^N$ we now clear up some of the confusions that were described by Jum Nunnally [N] and Jacob Cohen [Coh]:

Let $x_1, x_2, \dots$ be an infinite sample obtained by observation of independent and normally distributed real-valued random variables with parameters $(\theta, 1)$, where θ is an unknown mean and the variance is equal to 1. Using this infinite sample we want to estimate an unknown mean. If we denote by μ_θ a linear Gaussian measure on $\mathbf{R}$ with the probability density $\frac{1}{\sqrt{2\pi}}e^{-\frac{(x-\theta)^2}{2}}$, then the triplet

$$\big(\mathbf{R}^N, \mathbf{B}\big(\mathbf{R}^N\big), \mu_\theta^N\big)_{\theta \in R} \tag{5.1.1}$$

stands for a statistical structure described in our experiment, where $\mathbf{B}(\mathbf{R}^N)$ denotes the σ-algebra of Borel subsets of $\mathbf{R}^N$. By virtue of the strong law of large numbers we know that the condition

$$\mu_\theta^N\left(\left\{(x_k)_{k\in N} : (x_k)_{k\in N} \in \mathbf{R}^N \ \&\ \lim_{n\to\infty} \frac{\sum_{k=1}^n x_k}{n} = \theta\right\}\right) = 1 \tag{5.1.2}$$

G. Pantsulaia, *Applications of Measure Theory to Statistics*,
DOI 10.1007/978-3-319-45578-5_5

holds true for each $\theta \in \mathbf{R}$.

Taking into account the validity of (5.1.2), for construction of a consistent infinite sample estimation of an unknown parameter θ, as usual, a mapping T defined by

$$T((x_k)_{k\in N}) = \lim_{n\to\infty} \frac{\sum_{k=1}^{n} x_k}{n}, \tag{5.1.3}$$

is used in the theory of statistical decisions. It is well known that null hypothesis significance testing in the case $H_0 : \theta = \theta_0$ assumes the procedure: if an infinite sample $(x_k)_{k\in N} \in T^{-1}(\theta_0)$ then the H_0 hypothesis is accepted and the H_0 hypothesis is rejected, otherwise. There naturally arises a question asking whether Jacob Cohen's statement [Coh]: "... Don't look for a magic alternative to NHST [null hypothesis significance testing] ... It does not exist" can be explained. Note that a set S of all infinite samples $(x_k)_{k\in N}$ for which there exist finite limits of arithmetic means of their first n elements constitutes a proper Borel measurable vector subspace of $\mathbf{R}^N$. Following Christensen [Chr1], each proper Borel measurable vector subspace of an arbitrary Polish topological vector space is a Haar null set and inasmuch as S is a Borel measurable proper vector subspace of $\mathbf{R}^N$ we claim that the mapping T is not defined for "almost every" (in the sense of Christensen[1]) infinite sample. The latter relation means that for almost every infinite sample we reject the null hypothesis H_0. This discussion can also be used to explain Jum Nunnally's [N] conjecture.

If the decisions are based on convention they are termed arbitrary or mindless while those not so based may be termed subjective. To minimize type II errors, large samples are recommended. In psychology practically all null hypotheses are claimed to be false for sufficiently large samples so ...it is usually nonsensical to perform an experiment with the sole aim of rejecting the null hypothesis.

Now let $T_1 : \mathbf{R}^N \to R$ be another infinite sample consistent estimate of an unknown parameter θ in the above-mentioned model; that is,

$$\mu_\theta^N(\{(x_k)_{k\in N} : (x_k)_{k\in N} \in \mathbf{R}^N \;\&\; T_1((x_k)_{k\in N}) = \theta\}) = 1 \tag{5.1.4}$$

for each $\theta \in \mathbf{R}$. Here a question naturally arises about the additional conditions imposed on the estimate T_1 under which the above-described confusions will be settled.

In this direction, first note that there must be no parameter be set, because then for almost every infinite sample null hypothesis $H_0 : \theta = \theta_0$ will be rejected. Second, there must be no parameter $\theta_1 \in \mathbf{R}$ for which $T_1^{-1}(\theta_1)$ is a prevalent set (equivalently, a complement of a Haar null set) because then for almost every infinite sample null hypothesis $H_0 : \theta = \theta_2$ will be rejected for each $\theta_2 \neq \theta_1$. These observations lead us to additional conditions imposed on the estimate T_1 which assumes that $T_1^{-1}(\theta)$ must

[1]We say that a sentence $P(\cdot)$ formulated in terms of an element of a Polish group G is true for "almost every" element of G if a set of all elements $g \in G$ for which $P(g)$ is false constitutes a Haar null set in G.

be neither Haar null nor prevalent for each $\theta \in \mathbf{R}$. Following [BBE], a set which is neither Haar null nor prevalent is called a Haar ambivalent set. Such estimates first were adopted as objective infinite sample consistent estimates of a useful signal in the linear one-dimensional stochastic model (see, [ZPS], Theorem 4.1, p. 482).

It was proved in [ZPS] that $T_n : \mathbf{R}^n \to \mathbf{R}$ $(n \in N)$ defined by

$$T_n(x_1, \ldots, x_n) = -F^{-1}(n^{-1}\#(\{x_1, \ldots, x_n\} \cap (-\infty; 0])) \tag{5.1.5}$$

for $(x_1, \ldots, x_n) \in \mathbf{R}^n$, is a consistent estimator of a useful signal θ in the one-dimensional linear stochastic model

$$\xi_k = \theta + \Delta_k \ (k \in N), \tag{5.1.6}$$

where $\#(\cdot)$ denotes a counting measure, Δ_k is a sequence of independent identically distributed random variables on $\mathbf{R}$ with strictly increasing continuous distribution function F, and expectation of Δ_1 does not exist. The following two examples of simulations of a linear one-dimensional stochastic model (5.1.6) have been considered in [ZPS].

Example 5.1.1 ([ZPS], *Example 4.1, p. 484*) Because a sequence of real numbers $(\pi \times n - [\pi \times n])_{n \in N}$, where $[\cdot]$ denotes an integer part of a real number, is uniformly distributed on (0, 1) (see [KN], Example 2.1, p. 17), we claim that a simulation of a $\mu_{(\theta,1)}$ equidistributed sequence $(x_n)_{n \le M}$ on R (M is a "sufficiently large" natural number and depends on a representation quality of the irrational number π), where $\mu_{(\theta,1)}$ denotes a θ shift of the measure μ defined by distribution function F, can be obtained by the formula

$$x_n = F_\theta^{-1}(\pi \times n - [\pi \times n]) \tag{5.1.7}$$

for $n \le M$ and $\theta \in R$, where F_θ denotes a distribution function corresponding to the measure μ_θ.

In this model, θ stands for a "useful signal".

We set:

(i) n: The number of trials
(ii) T_n: An estimator defined by the formula (5.1.5)
(iii) $\overline{X}_n$: A sample average

When $F(x)$ is a standard Gaussian distribution function, by using Microsoft Excel we have obtained numerical data placed in Table 5.1.

Note that the results of computations presented in Table 5.1 show us that both statistics T_n and $\overline{X}_n$ give us good estimates of the useful signal θ whenever a generalized white noise in that case has a finite absolute moment of the first order, and its moment of the first order is equal to zero.

Now let F be a linear Cauchy distribution function on R; that is,

Table 5.1 Estimates of the useful signal $\theta = 1$ when the white noise is a standard Gaussian random variable

n	T_n	$\overline{X}_n$	n	T_n	$\overline{X}_n$
50	0.994457883	1.146952654	550	1.04034032	1.034899747
100	1.036433389	1.010190601	600	1.036433389	1.043940988
150	1.022241387	1.064790041	650	1.03313984	1.036321771
200	1.036433389	1.037987511	700	1.030325691	1.037905202
250	1.027893346	1.045296447	750	1.033578332	1.03728633
300	1.036433389	1.044049728	800	1.03108705	1.032630945
350	1.030325691	1.034339407	850	1.033913784	1.037321098
400	1.036433389	1.045181911	900	1.031679632	1.026202323
450	1.031679632	1.023083495	950	1.034178696	1.036669278
500	1.036433389	1.044635371	1000	1.036433389	1.031131694

Table 5.2 Estimates of the useful signal $\theta = 1$ when the white noise is a Cauchy random variable

n	T_n	$\overline{X}_n$	n	T_n	$\overline{X}_n$
50	1.20879235	2.555449288	550	1.017284476	41.08688757
100	0.939062506	1.331789564	600	1.042790358	41.30221291
150	1.06489184	71.87525566	650	1.014605804	38.1800532
200	1.00000000	54.09578271	700	1.027297114	38.03399768
250	1.06489184	64.59240343	750	1.012645994	35.57956117
300	1.021166379	54.03265563	800	1.015832638	35.25149408
350	1.027297114	56.39846672	850	1.018652839	33.28723503
400	1.031919949	49.58316089	900	1.0070058	31.4036155
450	1.0070058	44.00842613	950	1.023420701	31.27321466
500	1.038428014	45.14322051	1000	1.012645994	29.73405416

$$F(x) = \int_{-\infty}^{x} \frac{1}{\pi(1+t^2)} dt \ (x \in R). \tag{5.1.8}$$

Numerical data placed in Table 5.2 were obtained by using Microsoft Excel and a high-accuracy Cauchy distribution calculator [K].

On the one hand, the results of computations placed in Table 5.2 do not contradict the above-mentioned fact that T_n is a consistent estimator of the parameter $\theta = 1$. However, we know that a sample average $\overline{X}_n$ does not work in that case because the mean and variance of the white noise (i.e., Cauchy random variable) are not defined. For this reason attempts to estimate the useful signal $\theta = 1$ by using the sample average will not be successful.

In [ZPS] it has been established that the estimators $\overline{\lim}\widetilde{T_n} := \inf_n \sup_{m \geq n} \widetilde{T_m}$ and $\underline{\lim}\widetilde{T_n} := \sup_n \inf_{m \geq n} \widetilde{T_m}$ are consistent infinite sample estimates of a useful signal

θ in the model (5.1.6; see [ZPS], Theorem 4.2, p. 483). When we begin to study properties of these infinite sample estimators from the point of view of the theory of Haar null sets in $\mathbf{R}^N$, we observe a (for us) surprising and an unexpected fact that both these estimates are objective (see [PK3], Theorem 3.1).

As the described approach naturally divides a class of consistent infinite sample estimates into objective and subjective estimates should not seem excessively high showed our suggestion that each consistent infinite sample estimate must pass the theoretical test on objectivity.

The present chapter introduces the concepts of the theory of objective infinite sample consistent estimates in $\mathbf{R}^N$ and gives its extension to all nonlocally compact Polish groups admitting an invariant metric.

The rest of Chap. 5 is the following.

In Sect. 5.2 we give some notions and facts from the theory of Haar null sets in complete metric linear spaces and equidistributed sequences on the real axis $\mathbf{R}$. Concepts of objective and strong objective infinite sample consistent estimates for statistical structures are also introduced in this section. Section 5.3 presents a certain construction of the objective infinite sample consistent estimate of an unknown distribution function that generalizes the recent results obtained in [ZPS]. There we prove an existence of the infinite sample consistent estimate of an unknown distribution function $F(F \in \mathbf{F})$ for the family of Borel probability measures $\{p_F^N : F \in \mathbf{F}\}$, where $\mathbf{F}$ denotes the family of all strictly increasing and continuous distribution functions on $\mathbf{R}$ and p_F^N denotes an infinite power of the Borel probability measure p_F on $\mathbf{R}$ defined by F. Section 5.4 presents an effective construction of the strong objective infinite sample consistent estimate of the useful signal in a certain linear one-dimensional stochastic model. An infinite sample consistent estimate of an unknown probability density is constructed for the separated class of positive continuous probability densities and a problem about existence of an objective one is stated in Sect. 5.5. In Sect. 5.6, by using the notion of a Haar ambivalent set introduced in [BBE], essentially new classes of statistical structures having objective and strong objective estimates of an unknown parameter are introduced in a Polish nonlocally compact group admitting an invariant metric and relations between them are studied in this section. An example of a weakly separated statistical structure is constructed for which a question asking whether there exists a consistent estimate of an unknown parameter is not solvable within the theory (ZF) & (DC). These results extend recent results obtained in [PK2]. In addition, we extend the concept of objective and subjective consistent estimates introduced for $\mathbf{R}^N$ to all Polish groups and consider a question asking whether there exists an objective consistent estimate of an unknown parameter for any statistical structure in a nonlocally compact Polish group with an invariant metric when a subjective one exists. We show that this question is answered positively when there exists at least one such a parameter, the preimage of which under this subjective estimate is a prevalent. In Sect. 5.7 we consider some examples of objective and strong objective consistent estimates in a compact Polish group $\{0; 1\}^N$. In Sect. 5.8 we present a certain claim for theoretical statisticians in which each consistent estimation with domain in a nonlocally compact Polish group equipped with an invariant metric must pass the certification exam on the objectivity before its practical application and give some recommendations.

5.2 Auxiliary Notions and Facts from Functional Analysis and Measure Theory

Let V be a complete metric linear space, by which we mean a vector space (real or complex) with a complete metric for which the operations of addition and scalar multiplication are continuous. When we speak of a measure on V we always mean a nonnegative measure that is defined on the Borel sets of V and is not identically zero. We write $S + v$ for the translation of a set $S \subseteq V$ by a vector $v \in V$.

Definition 5.2.1 ([HSY], *Definition 1, p. 221*) A measure μ is said to be transverse to a Borel set $S \subset V$ if the following two conditions hold.

(i) There exists a compact set $U \subset V$ for which $0 < \mu(U) < 1$.
(ii) $\mu(S + v) = 0$ for every $v \in V$.

Definition 5.2.2 ([HSY], *Definition 2, p. 222*; [BBE], *p. 1579*) A Borel set $S \subset V$ is called shy if there exists a measure transverse to S. More generally, a subset of V is called shy if it is contained in a shy Borel set. The complement of a shy set is called a prevalent set. We say that a set is Haar ambivalent if it is neither shy nor prevalent.

Definition 5.2.3 ([HSY], *p. 226*) We say that "almost every" element of V satisfies some given property, if the subset of V on which this property holds is prevalent.

Lemma 5.2.1 ([HSY], Fact $3''$, p. 223) *The union of a countable collection of shy sets is shy.*

Lemma 5.2.2 ([HSY], Fact 8, p. 224) *If V is infinite-dimensional, all compact subsets of V are shy.*

Lemma 5.2.3 ([Kh], Lemma 2, p. 58) *Let μ be a Borel probability measure defined in complete separable metric space V. Then there exists a countable family of compact sets $(F_k)_{k\in N}$ in V such that $\mu(V \setminus \cup_{k\in N} F_k) = 0$.*

Let $\mathbf{R}^N$ be a topological vector space of all real-valued sequences equipped with Tychonoff metric ρ defined by $\rho((x_k)_{k\in N}, (y_k)_{k\in N}) = \sum_{k\in N} |x_k - y_k|/2^k(1 + |x_k - y_k|)$ for $(x_k)_{k\in N}, (y_k)_{k\in N} \in \mathbf{R}^N$.

Lemma 5.2.4 ([P2], Lemma 15.1.3, p. 202) *Let J be an arbitrary subset of N. We set*

$$A_J = \{(x_i)_{i\in N} : x_i \le 0 \, for \; i \in J \; \& \; x_i > 0 \, for \; i \in N \setminus J\}. \tag{5.2.1}$$

Then the family of subsets $\Phi = \{A_J : J \subseteq N\}$ has the following properties.

(i) Every element of Φ *is Haar ambivalent.*
(ii) $A_{J_1} \cap A_{J_2} = \emptyset$ *for all different* $J_1, J_2 \subseteq N$.
(iii) Φ *is a partition of* $\mathbf{R}^N$ *such that* $card(\Phi) = 2^{\aleph_0}$.

Remark 5.2.1 The proof of Lemma 5.2.4 employs an argument stating that each Borel subset of $\mathbf{R}^N$ which for each compact set contains its translate is a nonshy set.

Definition 5.2.4 ([KN]) A sequence $(x_k)_{k\in N}$ of real numbers from the interval (a, b) is said to be equidistributed or uniformly distributed on an interval (a, b) if for any subinterval $[c, d]$ of (a, b) we have

$$\lim_{n\to\infty} n^{-1}\#(\{x_1, x_2, \dots, x_n\} \cap [c, d]) = (b-a)^{-1}(d-c), \tag{5.2.2}$$

where # denotes a counting measure.

Now let X be a compact Polish space and μ be a probability Borel measure on X. Let $\mathbf{R}(X)$ be a space of all bounded continuous functions defined on X.

Definition 5.2.5 A sequence $(x_k)_{k\in N}$ of elements of X is said to be μ equidistributed or μ uniformly distributed on the X if for every $f \in \mathbf{R}(X)$ we have

$$\lim_{n\to\infty} n^{-1}\sum_{k=1}^{n} f(x_k) = \int_X f d\mu. \tag{5.2.3}$$

Lemma 5.2.5 ([KN], Lemma 2.1, p. 199) *Let* $f \in \mathbf{R}(X)$. *Then, for* μ^N *almost every sequence* $(x_k)_{k\in N} \in X^N$, *we have*

$$\lim_{n\to\infty} n^{-1}\sum_{k=1}^{n} f(x_k) = \int_X f d\mu. \tag{5.2.4}$$

Lemma 5.2.6 ([KN], pp. 199–201) *Let S be a set of all* μ *equidistributed sequences on X. Then we have* $\mu^N(S) = 1$.

Corollary 5.2.1 ([ZPS], Corollary 2.3, p. 473) *Let* ℓ_1 *be a Lebesgue measure on* $(0, 1)$. *Let D be a set of all* ℓ_1 *equidistributed sequences on* $(0, 1)$. *Then we have* $\ell_1^N(D) = 1$.

Definition 5.2.6 Let μ be a probability Borel measure on R and F be its corresponding distribution function. A sequence $(x_k)_{k\in N}$ of elements of R is said to be μ equidistributed or μ uniformly distributed on R if for every interval $[a, b](-\infty \le a < b \le +\infty)$ we have

$$\lim_{n\to\infty} n^{-1}\#([a, b] \cap \{x_1, \dots, x_n\}) = F(b) - F(a). \tag{5.2.5}$$

Lemma 5.2.7 ([ZPS], Lemma 2.4, p. 473) *Let $(x_k)_{k \in N}$ be an ℓ_1 equidistributed sequence on $(0, 1)$, F be a strictly increasing continuous distribution function on R, and p be a Borel probability measure on R defined by F. Then $(F^{-1}(x_k))_{k \in N}$ is p equidistributed on R.*

Remark 5.2.2 It can be shown that Definitions 5.2.5 and 5.2.6 are equivalent.

Corollary 5.2.2 ([ZPS], Corollary 2.4, p. 473) *Let F be a strictly increasing continuous distribution function on R and p_F be a Borel probability measure on R defined by F. Then for a set $D_F \subset R^N$ of all p equidistributed sequences on R we have:*
(i) $D_F = \{(F^{-1}(x_k))_{k \in N} : (x_k)_{k \in N} \in D\}$, where D comes from Corollary 5.2.1.
(ii) $p_F^N(D_F) = 1$.

Lemma 5.2.8 *Let F_1 and F_2 be two different strictly increasing continuous distribution functions on R and p_1 and p_2 be Borel probability measures on R defined by F_1 and F_2, respectively. Then there does not exist a sequence of real numbers $(x_k)_{k \in \mathbf{N}}$ that is simultaneously p_1 equidistributed and p_2 equidistributed.*

Theorem 5.2.1 *Let F_1 and F_2 be two different strictly increasing continuous distribution functions on R and p_1 and p_2 be Borel probability measures on R defined by F_1 and F_2, respectively. Then the measures p_1^N and p_2^N are orthogonal.*

Definition 5.2.7 Let $\{\mu_i : i \in I\}$ be a family of probability measures defined on a measure space (X, M). Let $L(I)$ denote a minimal σ-algebra generated by all singletons of I and $S(X)$ be the σ-algebra of subsets of X defined by

$$S(X) = \cap_{i \in I} dom(\overline{\mu}_i),$$

where $\overline{\mu}_i$ denotes the completion of the measure μ_i for $i \in I$.

We say that an $(S(X), L(I))$ measurable mapping $T : X \to I$ is a consistent (or well-founded) estimate of an unknown parameter i $(i \in I)$ for the family $\{\mu_i : i \in I\}$ if the condition

$$(\forall i)(i \in I \to \mu_i(T^{-1}(\{i\}) = 1)) \tag{5.2.6}$$

holds true.

Lemma 5.2.9 ([ZPS], Lemma 2.5, p. 474) *Let $\{\mu_i : i \in I\}$ be a family of probability measures defined on a measure space (X, M). The following sentences are equivalent.*
(i) The family of probability measures $\{\mu_i : i \in I\}$ is strongly separated.
(ii) There exists a consistent estimate of an unknown parameter i $(i \in I)$ for the family $\{\mu_i : i \in I\}$.

Now let $X_1, X_2, \ldots$ be an infinite sampling of independent, equally distributed real-valued random variables with unknown distribution function F. Assume that we know only that F belongs to the family of distribution functions $\{F_\theta : \theta \in \Theta\}$, where Θ is a nonempty set. Using these infinite samplings we want to estimate an unknown distribution function F. Let μ_θ denote a Borel probability measure on the real axis $\mathbf{R}$ generated by F_θ for $\theta \in \Theta$. We denote by μ_θ^N an infinite power of the measure μ_θ; that is, $\mu_\theta^N = \mu_\theta \times \mu_\theta \times \cdots$.

The triplet $(\mathbf{R}^N, \mathbf{B}(\mathbf{R}^N), \mu_\theta^N)_{\theta\in\Theta}$ is called a statistical structure describing our infinite experiment.

Definition 5.2.8 A Borel measurable function $T_n : \mathbf{R}^n \to \mathbf{R}$ $(n \in N)$ is called a consistent estimator of a parameter θ (in the sense of everywhere convergence) for the family $(\mu_\theta^N)_{\theta\in\Theta}$ if the condition

$$\mu_\theta^N(\{(x_k)_{k\in N} : (x_k)_{k\in N} \in \mathbf{R}^N \ \& \ \lim_{n\to\infty} T_n(x_1, \ldots, x_n) = \theta\}) = 1 \tag{5.2.7}$$

holds true for each $\theta \in \Theta$.

Definition 5.2.9 A Borel measurable function $T_n : \mathbf{R}^n \to \mathbf{R}$ $(n \in N)$ is called a consistent estimator of a parameter θ (in the sense of convergence in probability) for the family $(\mu_\theta^N)_{\theta\in\Theta}$ if for every $\epsilon > 0$ and $\theta \in \Theta$ the condition

$$\lim_{n\to\infty} \mu_\theta^N(\{(x_k)_{k\in N} : (x_k)_{k\in N} \in \mathbf{R}^N \ \& \ |T_n(x_1, \ldots, x_n) - \theta| > \epsilon\}) = 0 \tag{5.2.8}$$

holds true.

Definition 5.2.10 A Borel measurable function $T_n : \mathbf{R}^n \to \mathbf{R}$ $(n \in N)$ is called a consistent estimator of a parameter θ (in the sense of convergence in distribution) for the family $(\mu_\theta^N)_{\theta\in\Theta}$ if for every continuous bounded real-valued function f on $\mathbf{R}$ the condition

$$\lim_{n\to\infty} \int_{\mathbf{R}^N} f(T_n(x_1, \ldots, x_n)) d\mu_\theta^N((x_k)_{k\in N}) = f(\theta) \tag{5.2.9}$$

holds.

Remark 5.2.3 Following [Sh] (see Theorem 2, p. 272), for the family $(\mu_\theta^N)_{\theta\in R}$ we have:

(a) An existence of a consistent estimator of a parameter θ in the sense that everywhere convergence implies an existence of a consistent estimator of a parameter θ in the sense of convergence in probability

(b) An existence of a consistent estimator of a parameter θ in the sense of convergence in probability implies an existence of a consistent estimator of a parameter θ in the sense of convergence in distribution

Now let $L(\Theta)$ be a minimal σ-algebra of subsets generated by all singletons of the set Θ.

Definition 5.2.11 A $(\mathbf{B}(\mathbf{R}^N), L(\Theta))$-measurable function $T : \mathbf{R}^N \to \Theta$ is called an infinite sample consistent estimate (or estimator) of a parameter θ for the family $(\mu_\theta^N)_{\theta\in\Theta}$ if the condition

$$\mu_\theta^N(\{(x_k)_{k\in N} : (x_k)_{k\in N} \in \mathbf{R}^N \ \& \ T((x_k)_{k\in N}) = \theta\}) = 1 \tag{5.2.10}$$

holds true for each $\theta \in \Theta$.

Definition 5.2.12 An infinite sample consistent estimate $T : \mathbf{R}^N \to \Theta$ of a parameter θ for the family $(\mu_\theta^N)_{\theta\in\Theta}$ is called objective if $T^{-1}(\theta)$ is a Haar ambivalent set for each $\theta \in \Theta$. Otherwise, T is called subjective.

Definition 5.2.13 An objective infinite sample consistent estimate $T : \mathbf{R}^N \to \Theta$ of a parameter θ for the family $(\mu_\theta^N)_{\theta\in\Theta}$ is called strong if for each $\theta_1, \theta_2 \in \Theta$ there exists an isometric (with respect to the Tychonoff metric) transformation $A_{(\theta_1,\theta_2)}$ of $\mathbf{R}^N$ such that $A_{(\theta_1,\theta_2)}(T^{-1}(\theta_1)) \Delta T^{-1}(\theta_2)$ is shy.

Definition 5.2.14 Following [IS], the family $(\mu_\theta^N)_{\theta\in\Theta}$ is called strictly separated if there exists a family $(Z_\theta)_{\theta\in\Theta}$ of Borel subsets of $\mathbf{R}^N$ such that

(i) $\mu_\theta^N(Z_\theta) = 1$ for $\theta \in \Theta$.
(ii) $Z_{\theta_1} \cap Z_{\theta_2} = \emptyset$ for all different parameters θ_1 and θ_2 from Θ.
(iii) $\cup_{\theta\in\Theta} Z_\theta = \mathbf{R}^N$.

Remark 5.2.4 Note that an existence of an infinite sample consistent estimator of a parameter θ for the family $(\mu_\theta^N)_{\theta\in\Theta}$ implies that the family $(\mu_\theta^N)_{\theta\in\Theta}$ is strictly separated. Indeed, if we set $Z_\theta = \{(x_k)_{k\in N} : (x_k)_{k\in N} \in \mathbf{R}^N \ \& \ T((x_k)_{k\in N}) = \theta\}$ for $\theta \in \Theta$, then all conditions participating in Definition 5.2.14 will be satisfied.

Remark 5.2.5 Note that the test on objectivity for an infinite sample consistent estimate $T : \mathbf{R}^N \to \Theta$ of a parameter θ for the family $(\mu_\theta^N)_{\theta\in\Theta}$ is as follows. For each $\theta \in \Theta$, the set $T^{-1}(\theta)$ must be a Haar ambivalent set.

5.3 An Objective Infinite Sample Consistent Estimate of an Unknown Distribution Function

Theorem 5.3.1 *Let* $\mathbf{F}$ *be a family of distribution functions on* $\mathbf{R}$ *satisfying the following properties.*

(i) Each element of $\mathbf{F}$ *is strictly increasing and continuous.*
(ii) There exists a point x_* *such that* $F_1(x_*) \neq F_2(x_*)$ *for each different* $F_1, F_2 \in \mathbf{F}$.

Setting $\Theta = \{\theta = F(x_*) : F \in \mathbf{F}\}$ *and* $F_\theta = F$ *for* $\theta = F(x_*)$, *we get the following parameterization* $\mathbf{F} = \{F_\theta : \theta \in \Theta\}$. *We denote by* μ_θ *a Borel probability measure in* $\mathbf{R}$ *defined by* F_θ *for* $\theta \in \Theta$. *Then a function* $T_n : \mathbf{R}^n \to \mathbf{R}$, *defined by*

$$T_n(x_1, \dots, x_n) = \frac{\#(\{x_1, \dots, x_n\} \cap (-\infty; x_*])}{n} \tag{5.3.1}$$

for $(x_1, \dots, x_n) \in \mathbf{R}^n$ $(n \in N)$, *is a consistent estimator of a parameter* θ *for the family* $(\mu_\theta^N)_{\theta\in\Theta}$ *in the sense of almost everywhere convergence.*

Proof It is clear that T_n is Borel measurable function for $n \in N$. For $\theta \in \mathbf{R}$, we set

$$A_\theta = \{(x_k)_{k\in N} \ : \ (x_k)_{k\in N} \text{ is } \mu_\theta - equidistributed\ on\ \mathbf{R}\}. \tag{5.3.2}$$

Following Corollary 5.2.2, we have $\mu_\theta^N(A_\theta) = 1$ for $\theta \in \Theta$.

For $\theta \in \Theta$, we get

$$\mu_\theta^N(\{(x_k)_{k\in N} \in \mathbf{R}^N \ : \ \lim_{n\to\infty} T_n(x_1, \dots, x_n) = \theta\}) = \mu_\theta^N(\{(x_k)_{k\in N} \in \mathbf{R}^N \ :$$

$$\lim_{n\to\infty} n^{-1}\#(\{x_1, \dots, x_n\} \cap (-\infty; x_*]) = F_\theta(x_*)\}) \geq \mu_\theta^N(A_\theta) = 1. \tag{5.3.3}$$

The following corollaries are simple consequences of Theorem 5.3.1, Corollary 5.2.2, and Remarks 5.2.3 and 5.2.4.

Corollary 5.3.1 *An estimator T_n defined by (5.3.1) is a consistent estimator of a parameter θ for the family $(\mu_\theta^N)_{\theta\in\Theta}$ in the sense of convergence in probability.*

Corollary 5.3.2 *An estimator T_n defined by (5.3.1) is a consistent estimator of a parameter θ for the family $(\mu_\theta^N)_{\theta\in\Theta}$ in the sense of convergence in distribution.*

Theorem 5.3.2 *Let $\mathbf{F} = \{F_\theta : \theta \in \Theta\}$ and $(\mu_\theta^N)_{\theta\in\Theta}$ come from Theorem 5.3.1. Fix $\theta_0 \in \Theta$ and define an estimate $T_{\theta_0}^{(1)} : \mathbf{R}^N \to \Theta$ as follows. $T_{\theta_0}^{(1)}((x_k)_{k\in N}) = \overline{\lim}\widetilde{T}_n((x_k)_{k\in N})$ if $\overline{\lim}\widetilde{T}_n((x_k)_{k\in N}) \in \Theta \setminus \{\theta_0\}$ and $T_{\theta_0}^{(1)}((x_k)_{k\in N}) = \theta_0$, otherwise, where $\overline{\lim}\widetilde{T}_n = \inf_n \sup_{m\geq n} \widetilde{T}_m$ and*

$$\widetilde{T}_n((x_k)_{k\in N}) = n^{-1}\#(\{x_1, \dots, x_n\} \cap (-\infty; x_*])) \tag{5.3.4}$$

for $(x_k)_{k\in N} \in \mathbf{R}^N$. Then $T_{\theta_0}^{(1)}$ is an objective infinite sample consistent estimator of a parameter θ for the family $(\mu_\theta^N)_{\theta\in\Theta}$.

Proof Following [Sh] (see p. 189), the function $\overline{\lim}\widetilde{T}_n$ is Borel measurable which implies that the function $\overline{\lim}\widetilde{T}_n$ is $(\mathbf{B}(\mathbf{R}^N), L(\Theta))$ measurable. Following Corollary 5.2.2, we have $\mu_\theta^N(A_\theta) = 1$ for $\theta \in \Theta$, where A_θ is defined by (5.3.2). Therefore we get

$$\mu_\theta^N(\{(x_k)_{k\in N} \in \mathbf{R}^N \ : \ T_{\theta_0}^{(1)}(x_k)_{k\in N} = \theta\}) \geq \mu_\theta^N(\{(x_k)_{k\in N} \in \mathbf{R}^N \ : \ \overline{\lim}\widetilde{T}_n(x_k)_{k\in N} = \theta\})$$
$$\geq \mu_\theta^N(\{(x_k)_{k\in N} \in \mathbf{R}^N \ : \ \overline{\lim}\widetilde{T}_n(x_k)_{k\in N} = \underline{\lim}\widetilde{T}_n(x_k)_{k\in N} = F_\theta(x_*)\})$$
$$\geq \mu_\theta^N(A_\theta) = 1$$

for $\theta \in \Theta$.

Thus we have proved that the estimator $\mathbf{T}^{(1)}{}_{\theta_0}$ is an infinite sample consistent estimator of a parameter θ for the family $(\mu_\theta^N)_{\theta\in\Theta}$.

Now let us show that $\mathbf{T}^{(1)}{}_{\theta_0}$ is an objective infinite sample consistent estimator of a parameter θ for the family $(\mu_\theta^N)_{\theta\in\Theta}$.

Let us show that $B(\theta) := \left(\mathbf{T}^{(1)}{}_{\theta_0}\right)^{-1}(\theta)$ is a Haar ambivalent set for each $\theta \in \Theta$.

Let $(x_k)_{k\in N}$ be a μ_θ equidistributed sequence on $\mathbf{R}$. Then we get

$$\lim_{n\to\infty} n^{-1}\#(\{x_1,\ldots,x_n\}\cap(-\infty;x_*]) = \theta. \tag{5.3.5}$$

Let consider a set

$$C(\theta) = \{(y_k)_{k\in N} : y_k \le x_k \text{ if } x_k \le x_* \ \& \ y_k > x_k \text{ if } x_k > x_*\}. \tag{5.3.6}$$

Setting $J = \{k : x_k \le x_*\}$, we claim that $C(\theta) - (x_k)_{k\in N} = A_J$, where A_J comes from Lemma 5.2.4. Because any translate of a Haar ambivalent set is again a Haar ambivalent set, we claim that $C(\theta)$ is a Haar ambivalent set. A set $B(\theta)$ that contains the Haar ambivalent set $C(\theta)$ is nonshy. Because $\theta \in \Theta$ was taken arbitrarily we deduce that each B_θ is a Haar ambivalent set. The latter relation means that the estimator $\mathbf{T}^{(1)}{}_{\theta_0}$ is an objective infinite sample consistent estimator of a parameter θ for the family $(\mu_\theta^N)_{\theta\in\Theta}$.

Theorem 5.3.3 *Let* $\mathbf{F} = \{F_\theta : \theta \in \Theta\}$ *and* $(\mu_\theta^N)_{\theta\in\Theta}$ *come from Theorem 5.3.1. Fix* $\theta_0 \in \Theta$ *and define an estimate* $\mathbf{T}^{(2)}{}_{\theta_0} : \mathbf{R}^N \to \Theta$ *as follows.* $\mathbf{T}^{(2)}{}_{\theta_0}((x_k)_{k\in N}) = \underline{\lim}\widetilde{T}_n((x_k)_{k\in N})$ *if* $\underline{\lim}\widetilde{T}_n((x_k)_{k\in N}) \in \Theta \setminus \{\theta_0\}$ *and* $\mathbf{T}^{(2)}{}_{\theta_0}((x_k)_{k\in N}) = \theta_0$ *otherwise, where* $\underline{\lim}\widetilde{T}_n = \sup_n \inf_{m\ge n} \widetilde{T}_m$ *and*

$$\widetilde{T}_n((x_k)_{k\in N}) = n^{-1}\#(\{x_1,\ldots,x_n\}\cap(-\infty;x_*]) \tag{5.3.7}$$

for $(x_k)_{k\in N} \in \mathbf{R}^N$. *Then* $\mathbf{T}^{(2)}{}_{\theta_0}$ *is an objective infinite sample consistent estimator of a parameter* θ *for the family* $(\mu_\theta^N)_{\theta\in\Theta}$.

Proof Following [Sh] (see p. 189), the function $\underline{\lim}\widetilde{T}_n$ is Borel measurable which implies that the function $\overline{\lim}\widetilde{T}_n$ is $(\mathbf{B}(\mathbf{R}^n), L(\Theta))$ measurable. Following Corollary 5.2.2, we have $\mu_\theta^N(A_\theta) = 1$ for $\theta \in \Theta$, where A_θ is defined by (5.3.2). Thus we get

$$\begin{aligned}
&\mu_\theta^N(\{(x_k)_{k\in N} \in \mathbf{R}^N : \mathbf{T}^{(2)}{}_{\theta_0}(x_k)_{k\in N} = \theta\}) \ge \mu_\theta^N(\{(x_k)_{k\in N} \in \mathbf{R}^N : \underline{\lim}\widetilde{T}_n(x_k)_{k\in N} = \theta\})\\
&\ge \mu_\theta^N(\{(x_k)_{k\in N} \in \mathbf{R}^N : \overline{\lim}\widetilde{T}_n(x_k)_{k\in N} = \underline{\lim}\widetilde{T}_n(x_k)_{k\in N} = F_\theta(x_*)\})\\
&\ge \mu_\theta^N(A_\theta) = 1
\end{aligned}$$

for $\theta \in \Theta$.

Thus we have proved that the estimator $\mathbf{T}^{(2)}{}_{\theta_0}$ is an infinite sample consistent estimator of a parameter θ for the family $(\mu_\theta^N)_{\theta\in\Theta}$.

Now let us show that $\mathbf{T}^{(2)}{}_{\theta_0}$ is an objective infinite sample consistent estimator of a parameter θ for the family $(\mu_\theta^N)_{\theta\in\Theta}$.

Let us show that $B(\theta) = \left(\mathbf{T}^{(2)}{}_{\theta_0}\right)^{-1}(\theta)$ is a Haar ambivalent set for each $\theta \in \Theta$.

Let $(x_k)_{k\in N}$ be a μ_θ uniformly distributed sequence. Then we get

$$\lim_{n\to\infty} n^{-1}\#(\{x_1,\ldots,x_n\}\cap(-\infty;x_*])=\theta. \tag{5.3.8}$$

Let us consider a set

$$C(\theta)=\{(y_k)_{k\in N}:(y_k)_{k\in N}\in \mathbf{R}^N \;\&\; y_k\le x_k \text{ if } x_k\le x_* \;\&\; y_k>x_k \text{ if } x_k>x_*\}. \tag{5.3.9}$$

Setting $J=\{k : x_k\le x_*\}$, we deduce that $C(\theta)-(x_k)_{k\in N}=A_J$, where A_J comes from Lemma 5.2.4. Because any translate of a Haar ambivalent set is again a Haar ambivalent set, we claim that $C(\theta)$ is a Haar ambivalent set. A set $B(\theta)$ that contains the Haar ambivalent set $C(\theta)$ is nonshy. Because $\theta\in\Theta$ was taken arbitrary we deduce that each B_θ is a Haar ambivalent set. The latter relation means that the estimator $\mathbf{T}^{(2)}{}_{\theta_0}$ is an objective infinite sample consistent estimator of a parameter θ for the family $(\mu_\theta^N)_{\theta\in\Theta}$.

Remark 5.3.1 It can be shown that Theorems 5.3.2 and 5.3.3 extend the recent result obtained in [PK3] (see Theorem 3.1). Indeed, let us consider the linear one-dimensional stochastic system

$$(\xi_k)_{k\in N}=(\theta_k)_{k\in N}+(\Delta_k)_{k\in N}, \tag{5.3.10}$$

where $(\theta_k)_{k\in N}\in\mathbf{R}^N$ is a sequence of useful signals, $(\Delta_k)_{k\in N}$ is a sequence of independent identically distributed random variables (the so-called generalized "white noise") defined on some probability space $(\Omega,\mathbf{F},P)$ and $(\xi_k)_{k\in N}$ is a sequence of transformed signals. Let μ be a Borel probability measure on $\mathbf{R}$ defined by a random variable Δ_1. Then the N power of the measure μ denoted by μ^N coincides with the Borel probability measure on $\mathbf{R}^N$ defined by the generalized white noise; that is,

$$(\forall X)(X\in\mathbf{B}(\mathbf{R}^N)\to\mu^N(X)=P(\{\omega:\omega\in\Omega \;\&\; (\Delta_k(\omega))_{k\in N}\in X\})), \tag{5.3.11}$$

where $\mathbf{B}(\mathbf{R}^N)$ is the Borel σ-algebra of subsets of $\mathbf{R}^N$.

Following [IS], a general decision in information transmission theory is that the Borel probability measure λ, defined by the sequence of transformed signals $(\xi_k)_{k\in N}$ coincides with $\left(\mu^N\right)_{\theta_0}$ for some $\theta_0\in\Theta$ provided that

$$(\exists\theta_0)(\theta_0\in\Theta\to(\forall X)(X\in\mathbf{B}(\mathbf{R}^N)\to\lambda(X)=\left(\mu^N\right)_{\theta_0}(X))), \tag{5.3.12}$$

where $\left(\mu^N\right)_{\theta_0}(X)=\mu^N(X-\theta_0)$ for $X\in\mathbf{B}(\mathbf{R}^N)$.

Reference [ZPS] has considered a particular case of the above model (5.3.10) for which

$$(\theta_k)_{k\in N}\in\{(\theta,\theta,\ldots):\theta\in\mathbf{R}\}. \tag{5.3.13}$$

For $\theta\in\mathbf{R}$, a measure μ_θ^N defined by

$$\mu_\theta^N = \mu_\theta \times \mu_\theta \times \dots, \tag{5.3.14}$$

where μ_θ is a θ shift of μ (i.e., $\mu_\theta(X) = \mu(X - \theta)$ for $X \in \mathbf{B}(\mathbf{R})$), is called the N power of the θ shift of μ on $\mathbf{R}$.

Denote by F_θ a distribution function defined by μ_θ for $\theta \in \Theta$. Note that the family $\mathbf{F} = \{F_\theta : \theta \in \Theta\}$ satisfies all conditions participating in Theorem 5.3.1. Indeed, under x_* we can take the zero of the real axis. Then following Theorems 5.3.2 and 5.3.3, estimators $T^{(1)}_{\theta_0}$ and $T^{(2)}_{\theta_0}$ are objective infinite sample consistent estimators of a useful signal θ in the linear one-dimensional stochastic system (5.3.10). Note that these estimators exactly coincide with estimators constructed in [PK3] (see Theorem 5.3.1).

Theorem 5.3.4 *Let* $\mathbf{F}$ *be a family of all strictly increasing and continuous distribution functions in* $\mathbf{R}$ *and* p_F *be a Borel probability measure on* $\mathbf{R}$ *defined by* F *for each* $F \in \mathbf{F}$. *Then the family of Borel probability measures* $\{p_F^N : F \in \mathbf{F})\}$ *is strong separated.*

Remark 5.3.2 By virtue of the results of Lemma 5.2.9 and Theorem 5.3.4 we get that there exists a consistent estimate of an unknown distribution function F ($F \in \mathbf{F}$) for the family of Borel probability measures $\{p_F^N : F \in \mathbf{F}\}$, where $\mathbf{F}$ comes from Theorem 5.3.4. This estimate $T : \mathbf{R}^N \to \mathbf{F}$ is defined by $T((x_k)_{k\in N}) = F$ if $(x_k)_{k\in N} \in C_F$, where the family $(C_F)_{F\in\mathbf{F}}$ of subsets of $\mathbf{R}^N$ also comes from Theorem 4.4.2. Note that this result extends the main result established in [ZPS] (see Lemma 2.6, p. 476).

At end of this section we state the following.

Problem 5.3.2 Let $\mathbf{F}$ be a family of all strictly increasing and continuous distribution functions on $\mathbf{R}$ and p_F be a Borel probability measure in $\mathbf{R}$ defined by F for each $F \in \mathbf{F}$. Does there exist an objective infinite sample consistent estimate of an unknown distribution function F for the family of Borel probability measures $\{p_F^N : F \in \mathbf{F})\}$?

5.4 An Effective Construction of the Strong Objective Infinite Sample Consistent Estimate of a Useful Signal in the Linear One-Dimensional Stochastic Model

In [PK1], the examples of *objective* and *strong objective* infinite sample consistent estimates of a useful signal in the linear one-dimensional stochastic model were constructed by using the axiom of choice and a certain partition of the nonlocally compact Abelian Polish group R^N constructed in [P3].

In this section, in the same model we present an effective example of the strong objective infinite sample consistent estimate of a useful signal constructed in [PK2].

For each real number $a \in R$, we denote by $\{a\}$ its fractal part in the decimal system.

Theorem 5.4.1 *Let us consider the linear one-dimensional stochastic model (5.3.10), for which white noise has a infinite absolute moment of the first order and its moment of the first order is equal to zero. Suppose that the Borel probability measure* λ, *defined by the sequence of transformed signals* $(\xi_k)_{k\in N}$ *coincides with* $(\mu_{\theta_0}^N)$ *for some* $\theta_0 \in [0, 1]$. *Let* $T : R^N \to [0, 1]$ *be defined by:* $T((x_k)_{k\in N}) = \{\lim_{n\to\infty} \frac{\sum_{k=1}^n x_k}{n}\}$ *if* $\lim_{n\to\infty} \frac{\sum_{k=1}^n x_k}{n} \neq 1$, $T((x_k)_{k\in N}) = 1$ *if* $\lim_{n\to\infty} \frac{\sum_{k=1}^n x_k}{n} = 1$, *and* $T((x_k)_{k\in N}) = \sum_{k\in N} \frac{\chi_{(0,+\infty)}(x_k)}{2^k}$, *otherwise, where* $\chi_{(0,+\infty)}(\cdot)$ *denotes an indicator function of the set* $(0, +\infty)$ *defined on the real axis* R. *Then* T *is a strong objective infinite sample consistent estimate of the parameter* θ *for the statistical structure* $(R^N, \mathbf{B}(R^N), \mu_\theta^N)_{\theta\in\Theta}$ *describing the linear one-dimensional stochastic model (5.3.10).*

Proof **Step 1**. We have to show that T is an infinite sample consistent estimate of the parameter θ for the statistical structure $(R^N, \mathbf{B}(R^N), \mu_\theta^N)_{\theta\in\Theta}$ and $T^{-1}(\theta)$ is a Haar ambivalent set for each $\theta = \sum_{k=1}^\infty \frac{\theta_k}{2^k} \in \Theta$, where $\sum_{k=1}^\infty \frac{\theta_k}{2^k}$ is a representation of the number θ in the binary system.

Indeed, we have

$$(\forall\theta)(\theta \in (0, 1) \to T^{-1}(\theta) = (B_{H(\theta)} \setminus S) \cup \cup_{z\in Z} S_{\theta+z}), \tag{5.4.1}$$

where $H(\theta) = \{k : k \in N \ \&\ \theta_k = 1\}$, $B_{H(\theta)} = (\theta_k)_{k\in N} - A_{H(\theta)}$, $A_{H(\theta)}$ comes from Lemma 5.2.4,

$$S = \left\{(x_k)_{k\in N} \in \mathbf{R}^N : \text{exists a finite limit } \lim_{n\to\infty} \frac{\sum_{k=1}^n x_k}{n}\right\} \tag{5.4.2}$$

and

$$S_{\theta+z} = \left\{(x_k)_{k\in N} \in \mathbf{R}^N : \lim_{n\to\infty} \frac{\sum_{k=1}^n x_k}{n} = \theta + z\right\} \tag{5.4.3}$$

for each $\theta \in \Theta$ and $z \in Z$. Note that the set S like $\cup_{z\in Z} S_{\theta+z}$ is a Borel shy set (see [PK1], Lemma 4.14, p. 60). Taking into account this fact, the results of Lemma 5.2.4, invariance of Haar ambivalent sets under translations and symmetric transformation and the simple statement that difference of nonshy and shy sets is nonshy, we deduce that $T^{-1}(\theta)$ is a Borel measurable Haar ambivalent set for each $\theta \in \Theta$.

Note that

$$T^{-1}(1) = (B_{H(1)} \setminus S) \cup S_1 = (B_N \setminus S) \cup S_1 \tag{5.4.4}$$

and

$$T^{-1}(0) = (B_{H(0)} \setminus S) \cup \cup_{z\in Z\setminus\{1\}} S_{0+z} = (B_\emptyset \setminus S) \cup \cup_{z\in Z\setminus\{1\}} S_{0+z}, \tag{5.4.5}$$

which are also Borel measurable Haar ambivalent sets.

Now it is not hard to show that T is $(\mathbf{B}(R^N), L(\Theta))$ measurable because the class $\mathbf{B}(R^N)$ is closed under a countable family of set-theoretical operations and each element of $L(\Theta)$ is countable or cocountable in the interval $\Theta = [0, 1]$. Because $S_\theta \subseteq T^{-1}(\theta)$ for $\theta \in \Theta$, we deduce that $\mu_\theta(T^{-1}(\theta)) = 1$. The latter relation means that T is an infinite sample consistent estimate of a parameter θ.

Step 2. Let us show that for each different $\theta_1, \theta_2 \in [0, 1]$ there exists an isometric (with respect to the Tychonoff metric) transformation $A_{(\theta_1,\theta_2)}$ such that

$$A_{(\theta_1,\theta_2)}(T^{-1}(\theta_1)) \Delta T^{-1}(\theta_2) \tag{5.4.6}$$

is shy.

We define $A_{(\theta_1,\theta_2)}$ as follows. For $(x_k)_{k\in N} \in R^N$ we put $A_{(\theta_1,\theta_2)}((x_k)_{k\in N}) = (y_k)_{k\in N}$, where $y_k = -x_k$ if $k \in H(\theta_1) \Delta H(\theta_2) (:= (H(\theta_1) \setminus H(\theta_2)) \cup (H(\theta_2) \setminus H(\theta_1))$ and $y_k = x_k$ otherwise. It is obvious that $A_{(\theta_1,\theta_2)}$ is an isometric (with respect to the Tychonoff metric) transformation of the R^N.

Note that

$$A_{(\theta_1,\theta_2)}(T^{-1}(\theta_1)) \Delta T^{-1}(\theta_2) \subseteq \cup_{k\in N}\{0\}_k \times R^{N\setminus\{k\}} \cup S. \tag{5.4.7}$$

Both sets $\cup_{k\in N}\{0\}_k \times R^{N\setminus\{k\}}$ and S are shy, therefore by Lemma 5.2.1 and Definition 5.2.2 we claim that the set

$$A_{(\theta_1,\theta_2)}(T^{-1}(\theta_1)) \Delta T^{-1}(\theta_2) \tag{5.4.8}$$

is also shy.

This ends the proof of the theorem.

5.5 Infinite Sample Consistent Estimates of an Unknown Probability Density Function

Let $X_1, X_2, \dots$ be independent identically distributed real-valued random variables having a common probability density function f. After a so-called kernel class of estimates f_n of f based on $X_1, X_2, \dots, X_n$ was introduced by Rosenblatt [R], various convergence properties of these estimates have been studied. The stronger result in this direction was due to Nadaraya [N], who proved that if f is uniformly continuous then for a large class of kernels the estimates f_n converge uniformly on the real line to f with probability one. In [Sc] it has been shown that the above assumptions on f are necessary for this type of convergence. That is, if f_n converges uniformly to a function g with probability one, then g must be uniformly continuous and the distribution F from which we are sampling must be absolutely continuous with $F^{'}(x) = g(x)$ everywhere. When in addition to that mentioned above, it is assumed that f and its first $r + 1$ derivatives are bounded, it is possible to show how to construct estimates f_n such that $f_n^{(s)}$ converges uniformly to $f^{(s)}$ at a given rate with probability one for $s = 0, \dots, r$. Let $f_n(x)$ be a kernel estimate based on

$X_1, X_2, \ldots, X_n$ from F as given in [R]; that is,

$$f_n(x) = (na_n)^{-1} \sum_{i=1}^{n} k\left(\frac{X_i - x}{a_n}\right) \tag{5.5.1}$$

where $(a_n)_{n \in N}$ is a sequence of positive numbers converging to zero and k is a probability density function such that $\int_{-\infty}^{+\infty} |x|k(x)dx$ is finite and $k^{(s)}$ is a continuous function of bounded variation for $s = 0, \ldots, r$. The density function of the standard normal, for example, satisfies all these conditions.

In the sequel we need the following wonderful statement.

Lemma 5.5.1 ([Sc], Theorem 3.11, p. 1194) *A necessary and sufficient condition for*

$$\lim_{n \to \infty} \sup_{x \in \mathbf{R}} |f_n(x) - g(x)| = 0 \tag{5.5.2}$$

with probability one for a function g is that g be the uniformly continuous derivative of F.

Let $X_1, X_2, \ldots$ be independent and identically distributed real-valued random variables with an unknown probability density function f. Assume that we know that f belongs to the class of probability density function **SC**, each element of which is uniformly continuous.

Denote by $\ell^\infty(R)$ an infinite-dimensional nonseparable Banach space of all bounded real-valued functions on **R** equipped with norm $|| \cdot ||_\infty$ defined by

$$||h||_\infty = \sup_{x \in \mathbf{R}} |h(x)| \tag{5.5.3}$$

for all $h \in \ell^\infty(R)$. We say that $(\ell^\infty(R)) \lim_{n \to \infty} h_n = h_0$ if $\lim_{n \to \infty} ||h_n - h_0||_\infty = 0$.

Theorem 5.5.1 *Let ϕ denote a normal density function. We set $\Theta = \mathbf{SC}$. Let μ_θ be a Borel probability measure on* **R** *with probability density function $\theta \in \Theta$. Fix $\theta_0 \in \Theta$. For each $(x_i)_{i \in N}$ we set: $T_{\mathrm{SC}}((x_i)_{i \in N}) = (\ell^\infty(R)) \lim_{n \to \infty} f_n$ if this limit exists and is in $\Theta \setminus \{\theta_0\}$, and $T_{\mathrm{SC}}((x_i)_{i \in N}) = \theta_0$, otherwise. Then T_{SC} is a consistent infinite-sample estimate of an unknown parameter θ for the family $(\mu_\theta^N)_{\theta \in \Theta}$.*

Proof By Lemma 5.5.1, for each $\theta \in \Theta$ we have

$$\mu_\theta^N(\{(x_i)_{i \in N} \in \mathbf{R}^N : T_{\mathrm{SC}}((x_i)_{i \in N}) = \theta\}) \geq \mu_\theta^N(\{(x_i)_{i \in N} \in \mathbf{R}^N : (\ell^\infty(R)) \lim_{n \to \infty} f_n = \theta\})$$
$$= \mu_\theta^N(\{(x_i)_{i \in N} \in \mathbf{R}^N : \lim_{n \to \infty} ||f_n - \theta||_\infty = 0\}) = 1.$$

This ends the proof of the theorem.

Regarding Theorem 5.5.1 we state the following problems.

Problem 5.5.3 Let T_{SC} come from Theorem 5.5.1. Is T_{SC} an objective infinite sample consistent estimate of the parameter θ for the family $(\mu_\theta^N)_{\theta\in\Theta}$?

Problem 5.5.4 Let the statistical structure $\{(\mathbf{R}^N, \mathbf{B}(\mathbf{R}^N), \mu_\theta^N) : \theta \in \Theta\}$ come from Theorem 5.5.1. Does there exist an objective (or strong objective) infinite sample consistent estimate of the parameter θ for the family $(\mu_\theta^N)_{\theta\in\Theta}$?

Let $X_1, X_2, \dots$ be independent and identically distributed real-valued random variables with positive continuous probability density function f. Assume we know that f belongs to the separated class $\mathbf{A}$ of positive continuous probability densities provided there is a point x_* such that $g_1(x_*) \neq g_2(x_*)$ for each $g_1, g_2 \in \mathbf{A}$. Suppose we have an infinite sample $(x_k)_{k\in N}$ and we want to estimate an unknown probability density function. Setting $\Theta = \{\theta = g(x_*) : g \in \mathbf{A}\}$, we can give parameterization of the family $\mathbf{A}$ as follows: $\mathbf{A} = \{f_\theta : \theta \in \Theta\}$, where f_θ is a unique element f from the family $\mathbf{A}$ for which $f(x_*) = \theta$. Let μ_θ be a Borel probability measure defined by the probability density function f_θ for each $\theta \in \Theta$. It is obvious that $\{(\mathbf{R}^N, \mathbf{B}(\mathbf{R}^N), \mu_\theta^N) : \theta \in \Theta\}$ will be the statistical structure described in our experiment.

Theorem 5.5.2 *Let $(h_m)_{m\in N}$ be a sequence of a strictly decreasing sequence of positive numbers tending to zero. Let us fix $\theta_0 \in \Theta$. For each $(x_k)_{k\in N} \in \mathbf{R}^N$ we put*

$$T((x_k)_{k\in N}) = \lim_{m\to\infty}\lim_{n\to\infty} \frac{\#(\{x_1, \dots, x_n\} \cap [x_* - h_m, x_* + h_m])}{2nh_m} \tag{5.5.4}$$

if this repeated limit exists and belongs to the set $\Theta \setminus \{\theta\}$, and $T((x_k)_{k\in N}) = \theta_0$, otherwise. Then T is an infinite sample consistent estimate of the parameter θ for the family $(\mu_\theta^N)_{\theta\in\Theta}$.

Proof For each $\theta \in \Theta$, we put

$$A_\theta = \{(x_k)_{k\in N} : (x_k)_{k\in N} \in \mathbf{R}^N \ \&\ (x_k)_{k\in N} \text{ is } \mu_\theta - equidistributed\}. \tag{5.5.5}$$

By Corollary 5.2.2 we know that $\mu_\theta^N(A_\theta) = 1$ for each $\theta \in \Theta$.

For each $\theta \in \Theta$, we have

$$\mu_\theta^N(T^{-1}(\theta)) = \mu_\theta^N(\{(x_k)_{k\in N} \in \mathbf{R}^N \ :\ T((x_k)_{k\in N}) = \theta\})$$
$$\geq \mu_\theta^N(\{(x_k)_{k\in N} \in A_\theta \ :\ T((x_k)_{k\in N}) = \theta\})$$
$$= \mu_\theta^N\left(\left\{(x_k)_{k\in N} \in A_\theta \ :\ \lim_{m\to\infty} \frac{F_\theta(x_* + h_m) - F_\theta(x_* - h_m)}{2h_m} = \theta\right\}\right)$$
$$= \mu_\theta^N\left(\left\{(x_k)_{k\in N} \in A_\theta \ :\ \lim_{m\to\infty} \frac{\int_{x_*-h_m}^{x_*+h_m} f_\theta(x)dx}{2h_m} = \theta\right\}\right)$$
$$= \mu_\theta^N(\{(x_k)_{k\in N} \in A_\theta \ :\ f_\theta(x^*) = \theta\})$$
$$= \mu_\theta^N(A_\theta) = 1$$

Regarding Theorem 5.5.2 we state the following problems.

Problem 5.5.5 Let T come from Theorem 5.5.2. Is T an objective infinite sample consistent estimate of the parameter θ for the family $(\mu_\theta^N)_{\theta\in\Theta}$?

Problem 5.5.6 Let the statistical structure $\{(\mathbf{R}^N, \mathbf{B}(\mathbf{R}^N), \mu_\theta^N) : \theta \in \Theta\}$ come from Theorem 5.5.2. Does there exist an objective (or strong objective) infinite sample consistent estimate of the parameter θ for the family $(\mu_\theta^N)_{\theta\in\Theta}$?

Example 5.5.1 Let $X_1, X_2, \ldots$ be independent normally distributed real-valued random variables with parameters (a, σ) where a is a mean and σ is a standard deviation. Suppose that we know the mean a and want to estimate an unknown standard deviation σ by an infinite sample $(x_k)_{k\in N}$. For each $\sigma > 0$, denote by μ_σ the Gaussian probability measure on $\mathbf{R}$ with parameters (a, σ)(here $a \in \mathbf{R}$ is fixed). Let $(h_m)_{m\in N}$ be a sequence of a strictly decreasing sequence of positive numbers tending to zero.

By virtue of Theorem 5.5.2 we know that for each $\sigma > 0$ the following condition

$$\mu_\sigma^N\Bigg(\Bigg\{(x_k)_{k\in N} \in \mathbf{R}^N \ \&$$

$$\lim_{m\to\infty}\lim_{n\to\infty}\frac{\#(\{x_1,\ldots,x_n\}\cap[a-h_m, a+h_m])}{2nh_m} = \frac{1}{\sqrt{2\pi}\,\sigma}\Bigg\}\Bigg) = 1$$

holds true.

Fix $\sigma_0 > 0$. For $(x_k)_{k\in N} \in \mathbf{R}^N$ we put

$$T_1((x_k)_{k\in N}) = \lim_{m\to\infty}\lim_{n\to\infty}\frac{2nh_m}{\sqrt{2\pi}\#(\{x_1,\ldots,x_n\}\cap[a-h_m, a+h_m])} \tag{5.5.6}$$

if this limit exists and belongs to the set $(0, +\infty) \setminus \{\sigma_0\}$, and $T_1((x_k)_{k\in N}) = \sigma_0$, otherwise. Then for each $\sigma > 0$ we get

$$\mu_\sigma^N(\{(x_k)_{k\in N} : (x_k)_{k\in N} \in \mathbf{R}^N \ \& \ T_1((x_k)_{k\in N}) = \sigma\}) = 1 \tag{5.5.7}$$

which means that T_1 is an infinite sample consistent estimate of the standard deviation σ for the family $(\mu_\sigma^N)_{\sigma>0}$.

Theorem 5.5.3 *Let $X_1, X_2, \ldots$ be independent normally distributed real-valued random variables with parameters (a, σ), where a is a mean and σ is a standard deviation. Suppose that we know the mean a. Let $(a_n)_{n\in N}$ be a sequence of positive numbers converging to zero and ϕ be a standard Gaussian density function in* $\mathbf{R}$. *We denote by μ_σ a Borel Gaussian probability measure in $\mathbf{R}$ with parameters (a, σ) for each $\sigma \in \Sigma = (0, \infty)$. Fix $\sigma_0 \in \Sigma$ and define an estimate $\widetilde{T^{(1)}_{\sigma_0}} : \mathbf{R}^N \to \Sigma$ as follows. $\widetilde{T^{(1)}_{\sigma_0}}((x_k)_{k\in N}) = \overline{\lim}\widetilde{T^{(1)}_n}((x_k)_{k\in N})$ if $\overline{\lim}\widetilde{T^{(1)}_n}((x_k)_{k\in N}) \in \Sigma \setminus \{\sigma_0\}$ and $\widetilde{T^{(1)}_{\sigma_0}}((x_k)_{k\in N}) = \sigma_0$, otherwise, where $\overline{\lim}\widetilde{T^{(1)}_n} := \inf_n \sup_{m\geq n} \widetilde{T^{(1)}_m}$ and*

$$\widetilde{T_n^{(1)}}((x_k)_{k\in N}) = T_n^{(1)}(x_1,\dots,x_n) = \frac{1}{\sqrt{2\pi}(na_n)^{-1}\sum_{i=1}^n \phi(\frac{x_i-a}{a_n})} \tag{5.5.8}$$

for $(x_k)_{k\in N} \in \mathbf{R}^N$. *Then* $T_{\sigma_0}^{(1)}$ *is an infinite sample consistent estimator of a parameter* σ *for the family* $(\mu_\sigma^N)_{\sigma\in\Sigma}$.

Proof Following [Sh] (see p. 189), the function $\overline{\lim}\widetilde{T_n^{(1)}}$ is Borel measurable which implies that the function $\overline{\lim}\widetilde{T_n^{(1)}}$ is $(\mathbf{B}(\mathbf{R}^N), L(\Sigma))$ measurable.

For each $\sigma \in \Sigma$ we put

$$A_\sigma = \Big\{(x_k)_{k\in N} \in \mathbf{R}^N : lim_{n\to\infty}(na_n)^{-1}\sum_{i=1}^{n}\phi\Big(\frac{x_i-a}{a_n}\Big) = f_\sigma(a)\Big\}. \tag{5.5.9}$$

Because uniform convergence implies pointwise convergence, by Lemma 5.5.1 we deduce that $\mu_\sigma^N(A_\sigma) = 1$ for $\sigma \in \Sigma$ which implies

$$\mu_\sigma^N\Big(\Big\{(x_k)_{k\in N} \in \mathbf{R}^N \ : \ T_{\theta_0}^{(1)}(x_k)_{k\in N} = \sigma\Big\}\Big) \geq \mu_\sigma^N\Big(\Big\{(x_k)_{k\in N} \in \mathbf{R}^N \ : \ \overline{\lim}\widetilde{T_n^{(1)}}(x_k)_{k\in N} = \sigma\Big\}\Big)$$
$$\geq \mu_\sigma^N\Big(\Big\{(x_k)_{k\in N} \in \mathbf{R}^N \ : \ \overline{\lim}\widetilde{T_n^{(1)}}(x_k)_{k\in N} = \underline{\lim}\widetilde{T_n^{(1)}}(x_k)_{k\in N} = \sigma\Big\}\Big)$$
$$= \mu_\sigma^N\Big(\Big\{(x_k)_{k\in N} \in \mathbf{R}^N \ : \ \lim_{n\to\infty}\widetilde{T_n^{(1)}}((x_k)_{k\in N}) = \sigma\Big\}\Big)$$
$$= \mu_\sigma^N\Big(\Big\{(x_k)_{k\in N} \in \mathbf{R}^N \ : \ \lim_{n\to\infty}\frac{1}{\sqrt{2\pi}(na_n)^{-1}\sum_{i=1}^n \phi(\frac{x_i-a}{a_n})} = \sigma\Big\}\Big)$$
$$= \mu_\sigma^N\Big(\Big\{(x_k)_{k\in N} \in \mathbf{R}^N \ : \ \lim_{n\to\infty}(na_n)^{-1}\sum_{i=1}^{n}\phi\Big(\frac{x_i-a}{a_n}\Big) = \frac{1}{\sqrt{2\pi}\sigma}\Big\}\Big)$$
$$= \mu_\sigma^N\Big(\Big\{(x_k)_{k\in N} \in \mathbf{R}^N \ : \ \lim_{n\to\infty}(na_n)^{-1}\sum_{i=1}^{n}\phi\Big(\frac{x_i-a}{a_n}\Big) = f_\sigma(a)\Big\}\Big) = \mu_\sigma^N(A_\sigma) = 1. \tag{5.5.10}$$

The following theorem gives a construction of the objective infinite sample consistent estimate of an unknown parameter σ in the same model.

Theorem 5.5.4 *Let* $X_1, X_2, \dots$ *be independent normally distributed real-valued random variables with parameters* (a, σ), *where* a *is a mean and* σ *is a standard deviation. Suppose that we know the mean* a *is nonzero. Let* Φ *be a standard Gaussian distribution function in* $\mathbf{R}$. *We denote by* μ_σ *a Borel Gaussian probability measure in* $\mathbf{R}$ *with parameters* (a, σ) *for each* $\sigma \in \Sigma = (0, \infty)$. *Fix* $\sigma_0 \in \Sigma$ *and define an estimate* $T_{\sigma_0}^{(2)} : \mathbf{R}^N \to \Sigma$ *as follows.* $T_{\sigma_0}^{(2)}((x_k)_{k\in N}) = \overline{\lim}\widetilde{T_n^{(2)}}((x_k)_{k\in N})$ *if* $\overline{\lim}\widetilde{T_n^{(2)}}((x_k)_{k\in N}) \in \Sigma \setminus \{\sigma_0\}$ *and* $T_{\sigma_0}^{(2)}((x_k)_{k\in N}) = \sigma_0$, *otherwise, where* $\overline{\lim}\widetilde{T_n^{(2)}} := \inf_n \sup_{m\geq n}\widetilde{T_m^{(2)}}$ *and*

$$\widetilde{T_n^{(2)}}((x_k)_{k\in N}) = T_n^{(2)}(x_1,\dots,x_n) = -\frac{a}{\Phi^{-1}\big(\frac{\#(\{x_1,\dots,x_n\}\cap(-\infty,0])}{n}\big)} \tag{5.5.11}$$

for $(x_k)_{k\in N} \in \mathbf{R}^N$. *Then* $T^{(2)}_{\sigma_0}$ *is an objective infinite sample consistent estimator of a parameter* σ *for the family* $(\mu_\sigma^N)_{\sigma\in\Sigma}$.

Proof Following [Sh] (see p. 189), the function $\widetilde{\overline{\lim} T_n^{(2)}}$ is Borel measurable which implies that the function $\widetilde{\overline{\lim} T_n^{(2)}}$ is $(\mathbf{B}(\mathbf{R}^N), L(\Sigma))$ measurable.

For each $\sigma \in \Sigma$ we put

$$A_\sigma = \{(x_k)_{k\in N} \in \mathbf{R}^N : (x_k)_{k\in N} \text{ is } \mu_\sigma - \text{equidistributed in } \mathbf{R}\}. \tag{5.5.12}$$

By Corollary 5.2.16 we know that $\mu_\sigma^N(A_\sigma) = 1$ for $\sigma \in \Sigma$ which implies

$$\begin{aligned}
&\mu_\sigma^N\left(\left\{(x_k)_{k\in N} \in \mathbf{R}^N : \widetilde{\overline{\lim} T_n^{(2)}}(x_k)_{k\in N} = \sigma\right\}\right) \geq \mu_\sigma^N\left(\left\{(x_k)_{k\in N} \in \mathbf{R}^N : \widetilde{\overline{\lim} T_n^{(2)}}(x_k)_{k\in N}\right.\right.\\
&\left.\left.= \widetilde{\underline{\lim} T_n^{(2)}}(x_k)_{k\in N} = \sigma\right\}\right) = \mu_\sigma^N\left(\left\{(x_k)_{k\in N} \in \mathbf{R}^N : \lim_{n\to\infty} \widetilde{T_n^{(2)}}((x_k)_{k\in N}) = \sigma\right\}\right)\\
&= \mu_\sigma^N\left(\left\{(x_k)_{k\in N} \in \mathbf{R}^N : \lim_{n\to\infty} -\frac{a}{\Phi^{-1}\left(\frac{(\#(\{x_1,\dots,x_n\}\cap(-\infty,0])}{n}\right)} = \sigma\right\}\right)\\
&= \mu_\sigma^N\left(\left\{(x_k)_{k\in N} \in \mathbf{R}^N : \lim_{n\to\infty} \Phi^{-1}\left(\frac{(\#(\{x_1,\dots,x_n\}\cap(-\infty,0])}{n}\right) = -\frac{a}{\sigma}\right\}\right)\\
&= \mu_\sigma^N\left(\left\{(x_k)_{k\in N} \in \mathbf{R}^N : \lim_{n\to\infty} \frac{(\#(\{x_1,\dots,x_n\}\cap(-\infty,0])}{n} = \Phi\left(-\frac{a}{\sigma}\right)\right\}\right)\\
&= \mu_\sigma^N\left(\left\{(x_k)_{k\in N} \in \mathbf{R}^N : \lim_{n\to\infty} \frac{(\#(\{x_1,\dots,x_n\}\cap(-\infty,0])}{n} = \Phi_{(a,\sigma)}(0)\right\}\right)\\
&\geq \mu_\sigma^N(A_\sigma) = 1.
\end{aligned} \tag{5.5.13}$$

The latter relation means that $\widetilde{\overline{\lim} T_n^{(2)}}$ is a infinite sample consistent estimate of a parameter σ for the family of measures $(\mu_\sigma^N)_{\sigma>0}$.

Show that $\widetilde{\overline{\lim} T_n^{(2)}}$ is objective.

We have to show that for each $\sigma > 0$ the set $(\widetilde{\overline{\lim} T_n^{(2)}})^{-1}(\sigma)$ is a Haar ambivalent set.

Let $(x_k)_{k\in N}$ be a μ_σ equidistributed sequence. Then we get

$$\lim_{n\to\infty} \frac{(\#(\{x_1,\dots,x_n\}\cap(-\infty,0])}{n} = \Phi_{(a,\sigma)}(0) \tag{5.5.14}$$

which means

$$T^{(2)}_{\sigma_0}((x_k)_{k\in N}) = \widetilde{\overline{\lim} T_n^{(2)}}((x_k)_{k\in N}) = \sigma. \tag{5.5.15}$$

Setting $J_\sigma = \{i : x_i \leq 0\}$, it is not hard to show that a set

$$B_{J_\sigma} = \{(y_i)_{i\in N} : y_i \leq x_i \text{ for } i \in J_\sigma \;\&y_i > x_i \text{ for } i \in N \setminus J_\sigma\} \tag{5.5.16}$$

is a Haar ambivalent set.

It is also clear that for each $(y_i)_{i\in N} \in B_{J_\sigma}$ we have

$$T^{(2)}_{\sigma_0}((y_k)_{k\in N}) = \widetilde{\overline{\lim} T^{(2)}_n}((y_k)_{k\in N}) = \sigma,$$

which implies that $B_{J_\sigma} \subseteq (\widetilde{\overline{\lim} T^{(2)}_n})^{-1}(\sigma)$.

Because $\{(\widetilde{\overline{\lim} T^{(2)}_n})^{-1}(\sigma) : \sigma > 0\}$ is a partition of the $\mathbf{R}^N$ and each of them contains a Haar ambivalent set B_{J_σ} we deduce that $(\widetilde{\overline{\lim} T^{(2)}_n})^{-1}(\sigma)$ is a Haar ambivalent set for each $\sigma > 0$.

This ends the proof of the theorem.

Theorem 5.5.5 *Let $X_1, X_2, \ldots$ be independent normally distributed real-valued random variables with parameters (a, σ) where a is a mean and σ is a standard deviation. Suppose that both parameters are unknown. Let Φ be a standard Gaussian distribution function in $\mathbf{R}$. We denote by μ_σ a Borel Gaussian probability measure in $\mathbf{R}$ with parameters (a, σ) for each $\sigma \in \Sigma = (0, \infty)$ and $a \in \mathbf{R}$. Fix $\sigma_0 \in \Sigma$ and define an estimate $T^{(3)}_{\sigma_0} : \mathbf{R}^N \to \Sigma$ as follows. $T^{(3)}_{\sigma_0}((x_k)_{k\in N}) = \widetilde{\overline{\lim} T^{(3)}_n}((x_k)_{k\in N})$ if $\widetilde{\overline{\lim} T^{(3)}_n}((x_k)_{k\in N}) \in \Sigma \setminus \{\sigma_0\}$ and $T^{(3)}_{\sigma_0}((x_k)_{k\in N}) = \sigma_0$, otherwise, where $\overline{\lim} \widetilde{T^{(3)}_n} := \inf_n \sup_{m\geq n} \widetilde{T^{(3)}_m}$ and*

$$\widetilde{T^{(3)}_n}((x_k)_{k\in N}) = T^{(3)}_n(x_1, \ldots, x_n) = -\frac{\sum_{i=1}^n x_k}{n\Phi^{-1}\left(\frac{\#(\{x_1,\ldots,x_n\}\cap(-\infty,0])}{n}\right)} \tag{5.5.17}$$

for $(x_k)_{k\in N} \in \mathbf{R}^N$. Then $T^{(3)}_{\sigma_0}$ is an infinite sample consistent estimator of a parameter σ for the family $(\mu^N_\sigma)_{\sigma\in\Sigma}$.

Proof Following [Sh] (see p. 189), the function $\widetilde{\overline{\lim} T^{(3)}_n}$ is Borel measurable which implies that the function $\widetilde{\overline{\lim} T^{(3)}_n}$ is $(\mathbf{B}(\mathbf{R}^N), L(\Sigma))$ measurable.

For each $\sigma \in \Sigma$ we put

$$A_\sigma = \{(x_k)_{k\in N} \in \mathbf{R}^N : (x_k)_{k\in N} \text{ is } \mu_\sigma - \text{equidistributed in } \mathbf{R}\} \tag{5.5.18}$$

and

$$B_\sigma = \left\{(x_k)_{k\in N} \in \mathbf{R}^N : \lim_{n\to\infty} \frac{\sum_{k=1}^n x_k}{n} = a\right\}. \tag{5.5.19}$$

By Corollary 5.2.2 we know that $\mu^N_\sigma(A_\sigma) = 1$ for $\sigma \in \Sigma$. On the other hand, by the strong law of large numbers we know that $\mu^N_\sigma(B_\sigma) = 1$ for $\sigma \in \Sigma$. These relations imply that

$$\mu^N_\sigma(A_\sigma \cap B_\sigma) = 1 \tag{5.5.20}$$

for $\sigma \in \Sigma$.

Taking into account (5.5.20), we get

$$\mu_\sigma^N\Big(\Big\{(x_k)_{k\in N}\in \mathbf{R}^N \ :\ \widetilde{\overline{\lim} T_n^{(3)}}(x_k)_{k\in N}=\sigma\Big\}\Big)\geq \mu_\sigma^N\Big(\Big\{(x_k)_{k\in N}\in \mathbf{R}^N \ :\ \widetilde{\overline{\lim} T_n^{(3)}}(x_k)_{k\in N}$$
$$=\widetilde{\underline{\lim} T_n^{(3)}}(x_k)_{k\in N}=\sigma\Big\}\Big)=\mu_\sigma^N\Big(\Big\{(x_k)_{k\in N}\in \mathbf{R}^N \ :\ \lim_{n\to\infty}\widetilde{T_n^{(3)}}((x_k)_{k\in N})=\sigma\Big\}\Big)$$
$$=\mu_\sigma^N\Big(\Big\{(x_k)_{k\in N}\in \mathbf{R}^N \ :\ \lim_{n\to\infty}-\frac{\frac{\sum_{k=1}^n x_k}{n}}{\Phi^{-1}\big(\frac{(\#(\{x_1,\dots,x_n\}\cap(-\infty,0]))}{n}\big)}=\sigma\Big\}\Big)$$
$$\geq \mu_\sigma^N\Big(\Big\{(x_k)_{k\in N}\in A_\sigma\cap B_\sigma \ :\ \lim_{n\to\infty}\Phi^{-1}\Big(\frac{(\#(\{x_1,\dots,x_n\}\cap(-\infty,0])}{n}\Big)$$
$$=-\lim_{n\to\infty}\frac{\frac{\sum_{k=1}^n x_k}{n}}{\sigma}\Big\}\Big)=\mu_\sigma^N\Big(\Big\{(x_k)_{k\in N}\in A_\sigma\cap B_\sigma \ :\ \lim_{n\to\infty}\Phi^{-1}\Big(\frac{(\#(\{x_1,\dots,x_n\}\cap(-\infty,0])}{n}\Big)$$
$$=-\frac{a}{\sigma}\Big\}\Big)=\mu_\sigma^N\Big(\Big\{(x_k)_{k\in N}\in \mathbf{R}^N \ :\ \lim_{n\to\infty}\frac{(\#(\{x_1,\dots,x_n\}\cap(-\infty,0])}{n}=\Phi_{(a,\sigma)}(0)\Big\}\Big)$$
$$=\mu_\sigma^N(A_\sigma\cap B_\sigma)=1. \tag{5.5.21}$$

The latter relation means that $T_{\sigma_0}^{(3)}$ is an infinite sample consistent estimate of a parameter σ for the family of measure s $(\mu_\sigma^N)_{\sigma>0}$.

This ends the proof of the theorem.

Example 5.5.2 Because a sequence of real numbers $(\pi\times n-[\pi\times n])_{n\in N}$, where $[\cdot]$ denotes an integer part of a real number, is uniformly distributed on $(0,1)$(see [KN], Example 2.1, p. 17), we claim that a simulation of a $\mu_{(3,5)}$ equidistributed sequence $(x_n)_{n\leq M}$ on R (M is a "sufficiently large" natural number and depends on a representation quality of the irrational number π), where $\mu_{(3,5)}$ denotes a linear Gaussian measure with parameters (3, 5), can be obtained by the formula

$$x_n=\Phi_{(3,5)}^{-1}(\pi\times n-[\pi\times n]) \tag{5.5.22}$$

for $n\leq M$, where $\Phi_{(3,5)}$ denotes a Gaussian distribution function with parameters $(3,5)$.

Suppose that $a=3$ and we want to estimate an unknown standard deviation σ.

We set: n, the number of trials; S_n, a square root from the sample variance; $S_n^{'}$, a square root from the corrected sample variance; $T_n^{(1)}$, an estimate defined by the formula (5.5.11); $T_n^{(3)}$, an estimate defined by the formula (5.5.17); and σ, an unknown standard deviation.

The numerical data placed in Table 5.3 were obtained by using Microsoft Excel.

Note that the results of computations presented in Table 5.3 show us that both statistics $T_n^{(1)}$ and $T_n^{(3)}$ work correctly.

At end of this section we state the following.

Problem 5.5.7 Let $\mathbf{D}$ be a class of positive continuous probability densities and p_f be a Borel probability measure on $\mathbf{R}$ with probability density function f for each

Table 5.3 Estimates of an unknown standard deviation $\sigma = 5$

n	S_n	$S_n^{'}$	$T_n^{(1)}$	$T_n^{(3)}$
200	4.992413159	5.004941192	5.205401325	4.895457577
400	4.992413159	5.004941192	5.141812921	4.835655399
600	5.10523925	5.109498942	5.211046737	4.855457413
800	5.106390271	5.109584761	5.19369988	4.92581015
1000	5.066642282	5.069177505	5.028142523	4.944169095
1200	5.072294934	5.074409712	5.235885276	4.935995814
1400	5.081110418	5.082926073	5.249446371	4.96528786
1600	5.079219075	5.080807075	5.205452797	4.9564705
1800	5.060850283	5.06225666	5.207913228	4.963326232
2000	5.063112113	5.064378366	5.239119585	4.981223889

$f \in \mathbf{D}$. Does there exist an objective (or a subjective) infinite sample consistent estimate of an unknown probability density function f for the family of Borel probability measures $\{p_f^N : f \in \mathbf{D})\}$?

5.6 Orthogonal Statistical Structures in a Nonlocally Compact Polish Group Admitting an Invariant Metric

Let G be a Polish group, by which we mean a separable group with a complete invariant metric ρ (i.e., $\rho(fh_1g, fh_2g) = \rho(h_1, h_2)$ for each $f, g, h_1, h_2 \in G$) for which the transformation (from $G \times G$ onto G) that sends (x, y) into $x^{-1}y$, is continuous. Let $\mathbf{B}(G)$ denote the σ-algebra of Borel subsets of G.

Definition 5.6.1 ([M]) A Borel set $X \subseteq G$ is called shy, if there exists a Borel probability measure μ over G such that $\mu(fXg) = 0$ for all $f, g \in G$. A measure μ is called a testing measure for a set X. A subset of a Borel shy set is also called shy. The complement of a shy set is called a prevalent set.

Definition 5.6.2 ([BBE]) A Borel set is called a Haar ambivalent set if it is neither shy nor prevalent.

Remark 5.6.1 Note that if $X \subseteq G$ is shy then there exists a testing measure μ for a set X that has a compact carrier $K \subseteq G$(i.e., $\mu(G \setminus K) = 0$). The collection of shy sets constitutes the σ ideal, and in the case when G is locally compact, a set is shy iff it has Haar measure zero.

Definition 5.6.3 If G is a Polish group and $\{\mu_\theta : \theta \in \Theta\}$ is a family of Borel probability measures on G, then the family of triplets $\{(G, \mathbf{B}, \mu_\theta) : \theta \in \Theta\}$, where Θ is a nonempty set equipped with the σ-algebra $L(\Theta)$ generated by all singletons of Θ, is called a statistical structure. A set Θ is called a set of parameters.

Definition 5.6.4 (**O**) The statistical structure $\{(G, \mathbf{B}(G), \mu_\theta) : \theta \in \Theta\}$ is called orthogonal if the measures μ_{θ_1} and μ_{θ_2} are orthogonal for each different parameter θ_1 and θ_2.

Definition 5.6.5 (**WS**) The statistical structure $\{(G, \mathbf{B}(G), \mu_\theta) : \theta \in \Theta\}$ is called weakly separated if there exists a family of Borel subsets $\{X_\theta : \theta \in \Theta\}$ such that $\mu_{\theta_1}(X_{\theta_2}) = \delta(\theta_1, \theta_2)$, where δ denotes Kronecker's function defined on the Cartesian square $\Theta \times \Theta$ of the set Θ.

Definition 5.6.6 (**SS**) The statistical structure $\{(G, \mathbf{B}(G), \mu_\theta) : \theta \in \Theta\}$ is called strong separated (or strictly separated) if there exists a partition of the group G into the family of Borel subsets $\{X_\theta : \theta \in \Theta\}$ such that $\mu_\theta(X_\theta) = 1$ for each $\theta \in \Theta$.

Definition 5.6.7 (**CE**) A $(\mathbf{B}(G), L(\Theta))$ measurable mapping $T : G \to \Theta$ is called a consistent estimate of an unknown parameter $\theta \in \Theta$ for the statistical structure $\{(G, \mathbf{B}(G), \mu_\theta) : \theta \in \Theta\}$ if the condition $\mu_\theta(T^{-1}(\theta)) = 1$ holds true for each $\theta \in \Theta$.

Definition 5.6.8 (**OCE**) A $(\mathbf{B}(G), L(\Theta))$ measurable mapping $T : G \to \Theta$ is called an objective consistent estimate of an unknown parameter $\theta \in \Theta$ for the statistical structure $\{(G, \mathbf{B}(G), \mu_\theta) : \theta \in \Theta\}$ if the following two conditions hold.
(i) $\mu_\theta(T^{-1}(\theta)) = 1$ for each $\theta \in \Theta$.
(ii) $T^{-1}(\theta)$ is a Haar ambivalent set for each $\theta \in \Theta$.

If Condition (i) holds but Condition (ii) fails, then T is called a subjective consistent estimate of an unknown parameter $\theta \in \Theta$ for the statistical structure $\{(G, \mathbf{B}, \mu_\theta) : \theta \in \Theta\}$.

Definition 5.6.9 (**SOCE**) An objective consistent estimate $T : G \to \Theta$ of an unknown parameter $\theta \in \Theta$ for the statistical structure $\{(G, \mathbf{B}(G), \mu_\theta) : \theta \in \Theta\}$ is called *strong* if for each $\theta_1, \theta_2 \in \Theta$ there exists an isometric Borel measurable bijection $A_{(\theta_1,\theta_2)} : G \to G$ such that the set $A_{(\theta_1,\theta_2)}(T^{-1}(\theta_1)) \Delta T^{-1}(\theta_2)$ is shy in G.

Remark 5.6.2 Let G be a Polish nonlocally compact group admitting an invariant metric. The relations between statistical structures introduced in Definitions 5.6.4–5.6.9 for such a group can be represented by the following diagram.

$$\mathbf{SOCE} \to \mathbf{OCE} \to \mathbf{CE} \leftrightarrow \mathbf{SS} \to \mathbf{WS} \to \mathbf{O} \tag{5.6.1}$$

To show that the converse implications sometimes fail we consider the following examples.

Example 5.6.1 $\rceil$(**WS** $\leftarrow$ **O**) Let $F \subset G$ be a closed subset of the cardinality $2^{\aleph_0}$. Let $\phi : [0, 1] \to F$ be a Borel isomorphism of $[0, 1]$ onto F. We set $\mu(X) = \lambda(\phi^{-1}(X \cap F))$ for $X \in \mathbf{B}(G)$, where λ denotes a linear Lebesgue measure on $[0, 1]$. We put $\Theta = F$. Fix $\theta_0 \in \Theta$ and put $\mu_\theta = \mu$ if $\theta = \theta_0$, and $\mu_\theta = \delta_\theta|_{\mathbf{B}(G)}$, otherwise, where δ_θ denotes a Dirac measure on G concentrated at the point θ and $\delta_\theta|_{\mathbf{B}(G)}$ denotes the restriction of the δ_θ to the class $\mathbf{B}(G)$. Then the statistical structure

$\{(G, \mathbf{B}, \mu_\theta) : \theta \in \Theta\}$ stands for **O** which is not **WS**. Indeed, if we assume there is a family of Borel subsets $\{X_\theta : \theta \in \Theta\}$ for which $\mu_{\theta_1}(X_{\theta_2}) = \delta(\theta_1, \theta_2)$ for each $\theta_1, \theta_2 \in \Theta$, then we get that $\mu_\theta(X_{\theta_0}) = 0$ for each $\theta \in \Theta \setminus \{\theta_0\}$. The latter relation means that $\{\theta\} \cap X_{\theta_0} = \emptyset$ for each $\theta \in \Theta \setminus \{\theta_0\}$. Hence, $\Theta \setminus \{\theta_0\}\{\theta\} \cap X_{\theta_0} = \emptyset$ which implies that $\mu_{\theta_0}(X_{\theta_0}) = 0$ because $\Theta \setminus \{\theta_0\}$ is a carrier of μ_{θ_0} and its intersection with the set X_{θ_0} is emptyset.

Example 5.6.2 (*SM*) $\rceil$(**SS** $\leftarrow$ **WS**) Following [P1] (see Theorem 1, p. 335), in the system of axioms (ZFC) the following three conditions are equivalent.

(1) The continuum hypothesis ($c = 2^{\aleph_0} = \aleph_1$).
(2) For an arbitrary probability space $(E; S; \mu)$, the μ measure of the union of any family $(E_i)_{i \in I}$ of μ measure zero subsets, such that $\mathrm{card}(I) < c$, is equal to zero.
(3) An arbitrary weakly separated family of probability measures, of cardinality continuum, is strictly separated.

The latter relation means that under the continuum hypothesis in ZFC we have **SS** $\leftarrow$ **WS**. This is just Skorohod's well-known result (see [IS]). Moreover, following [P1] (see Theorem 2, p. 339), if (F, ρ) is a Radon metric space and $(\mu_i)_{i \in I}$ is a weakly separated family of Borel probability measures with $\mathrm{card}(I) \leq c$, then in the system of axioms $(ZFC)\&(MA)$, the family $(\mu_i)_{i \in I}$ is strictly separated.

Let us consider a counterexample to the implication **SS** $\leftarrow$ **WS** in the Solovay model (SM) [Solovay] which is the following system of axioms. $(ZF) + DC$ +"every subset of the real axis **R** is Lebesgue measurable," where (ZF) denotes the Zermelo–Fraenkel set theory and (DC) denotes the axiom of dependence choices.

For $\theta \in (0; 1)$, let b_θ be a linear classical Borel measure defined on the set $\{\theta\} \times (0; 1)$. For $\theta \in (1.2)$, let b_θ be a linear classical Borel measure defined on the set $(0; 1) \times \{\theta - 1\}$. By λ_θ we denote a Borel probability measure on $(0; 1) \times (0; 1)$ produced by b_θ; that is,

$$(\forall X)(\forall \theta_1)(\forall \theta_2)(X \in \mathbf{B}((0; 1) \times (0; 1)) \;\&\; \theta_1 \in (0; 1) \;\&\; \theta_2 \in (1; 2) \rightarrow$$

$$\lambda_{\theta_1}(X) = b_{\theta_1}((\{\theta_1\} \times (0; 1)) \cap X) \;\&\; \lambda_{\theta_2}(X) = b_{\theta_2}(((0; 1) \times \{\theta_1 - 1\}) \cap X)). \quad (5.6.2)$$

If we put $\theta = (0; 1) \cup (1; 2)$, then we get a statistical structure

$$((0; 1) \times (0; 1), \mathbf{B}((0; 1) \times (0; 1)), \lambda_\theta)_{\theta \in \Theta}. \quad (5.6.3)$$

Setting $X_\theta = \{\theta\} \times (0; 1)$ for $\theta \in (0; 1)$, and $X_\theta = (0; 1) \times \{\theta - 1\}$ for $\theta \in (1.2)$, we observe that for the family of Borel subsets $\{X_\theta : \theta \in \Theta\}$ we have $\lambda_{\theta_1}(X_{\theta_2}) = \delta(\theta_1, \theta_2)$, where δ denotes Kronecker's function defined on the Cartesian square $\Theta \times \Theta$ of the set Θ. In other words, $(\lambda_\theta)_{\theta \in \Theta}$ is weakly separated. Now let us assume that this family is strong separated. Then there will be a partition $\{Y_\theta : \theta \in \Theta\}$ of the set $(0; 1) \times (0; 1)$ into Borel subsets $(Y_\theta)_{\theta \in \Theta}$ such that $\lambda_\theta(Y_\theta) = 1$ for each $\theta \in \Theta$. If we consider $A = \cup_{\theta \in (0;1)} Y_\theta$ and $B = \cup_{\theta \in (1;2)} Y_\theta$ then we observe by the Fubini theorem that $\ell_2(A) = 1$ and $\ell_2(B) = 1$, where ℓ_2 denotes the 2-dimensional

Lebesgue measure defined in $(0; 1) \times (0; 1)$. This is the contradiction and we proved that $(\lambda_\theta)_{\theta\in\Theta}$ is not strictly separated. An existence of a Borel isomorphism g between $(0; 1) \times (0; 1)$ and G allows us to construct a family $(\mu_\theta)_{\theta\in\Theta}$ in G: $\mu_\theta(X) = \lambda_\theta(g^{-1}(X))$ for each $X \in \mathbf{B}(G)$ and $\theta \in \Theta$ which is **WS** but not **SS**(equivalently, **CE**). By virtue of the celebrated result of Mycielski and Swierczkowski (see [MS]) who asserted that under the axiom of determinacy (AD) every subset of the real axis $\mathbf{R}$ is Lebesgue measurable, the same example can be used as a counterexample to the implication **SS** $\leftarrow$ **WS** in the theory $(ZF) + (DC) + (AD)$. The answer to the question asking *whether* $(\mu_\theta)_{\theta\in\Theta}$ *has a consistent estimate* is **yes** in the theory (ZFC) & (CH), and **no** in the theory $(ZF) + (DC) + (AD)$, we deduce that this question is not solvable within the theory $(ZF) + (DC)$.

Example 5.6.3 $\rceil$(**OCE** $\leftarrow$ **CE**) Setting $\Theta = G$ and $\mu_\theta = \delta_\theta|\mathbf{B}(G)$ for $\theta \in \Theta$, where δ_θ denotes a Dirac measure in G concentrated at the point θ and $\delta_\theta|\mathbf{B}(G)$ denotes its restriction to $\mathbf{B}(G)$, we get a statistical structure $(G, \mathbf{B}(G), \mu_\theta)_{\theta\in\Theta}$. Let $L(\Theta)$ denote a minimal σ-algebra of subsets of Θ generated by all singletons of Θ. Setting $T(g) = g$ for $g \in G$, we get a consistent estimate of an unknown parameter θ for the family $(\mu_\theta)_{\theta\in\Theta}$. Note that there does not exist an objective consistent estimate of a parameter θ for the family $(\mu_\theta)_{\theta\in\Theta}$. Indeed, if we assume the contrary and that T_1 is such an estimate, we get that $T_1^{-1}(\theta)$ is a Haar ambivalent set for each $\theta \in \Theta$. Because T_1 is a consistent estimate of an unknown parameter θ for each $\theta \in \Theta$, we get that the condition $\mu_\theta(T_1^{-1}(\theta)) = 1$ holds true which implies that $\theta \in T_1^{-1}(\theta)$ for each $\theta \in \Theta$. Fix any parameter $\theta_0 \in \Theta$. Because $T_1^{-1}(\theta_0)$ is a Haar ambivalent set there is $\theta_1 \in T_1^{-1}(\theta_0)$ which differs from θ_0. Then $T_1^{-1}(\theta_0)$ and $T_1^{-1}(\theta_1)$ are not disjoint because $\theta_1 \in T_1^{-1}(\theta_0) \cap T_1^{-1}(\theta_1)$ and we get the contradiction.

Remark 5.6.3 Note that if (Θ, ρ) is a metric space and if in Definition 5.6.7 the requirement of a $(\mathbf{B}(G), L(\Theta))$ measurability is replaced with a $(\mathbf{B}(G), \mathbf{B}(\Theta))$ measurability, then the implication **SS** $\rightarrow$ **CE** may be false. Indeed, let G be a Polish group and $f : G \leftarrow \Theta(:= G)$ be a nonmeasurable(in the Borel sense) bijection. For each $\theta \in \Theta$ denote by μ_θ the restriction of the Dirac measure $\delta_{f(\theta)}$ to the σ-algebra of Borel subsets of the group G. It is clear that the statistical structure $\{(G, \mathbf{B}(G), \mu_\theta) : \theta \in \Theta\}$ is strictly separated. Show that a consistent estimate for that statistical structure does not exist. Indeed, let $T : G \rightarrow \Theta$ be a $(\mathbf{B}(G), \mathbf{B}(\Theta))$ measurable mapping such that $\mu_\theta(\{x : T(x) = \theta\}) = 1$ for each $\theta \in \Theta$. Because the measure μ_θ is concentrated at the point $f(\theta)$ we have that $f(\theta) \in \{x : T(x) = \theta\}$ for each $\theta \in \Theta$ which implies that $T(f(\theta)) = \theta$ for each $\theta \in \Theta$. The latter relation means that $T = f^{-1}$. Inasmuch as f is not $(\mathbf{B}(G), \mathbf{B}(\Theta))$ measurable, we claim that $f^{-1} = T$ is also not $(\mathbf{B}(G), \mathbf{B}(\Theta))$ measurable and we get the contradiction.

There naturally arises a question of whether there exists a statistical structure $\{(G, \mathbf{B}, \mu_\theta) : \theta \in \Theta\}$ in a Polish nonlocally compact group admitting an invariant metric that has an objective consistent estimate of a parameter θ. To answer this question positively, we need the following two lemmas.

Lemma 5.6.1 ([Sol], Theorem, p. 206) *Assume G is a Polish, nonlocally compact group admitting an invariant metric. Then there exists a closed set $F \subseteq G$ and a* *continuous function* $\phi : F \to 2^N$ *such that for any* $x \in 2^N$ *and any compact set* $K \subseteq G$ *there is* $g \in G$ *with* $gK \subseteq \phi^{-1}(x)$.

Lemma 5.6.2 ([D], Proposition 12, p. 87) *Let G be a nonlocally compact Polish group with an invariant metric. Then any compact subset (and hence any* K_σ *subset) of G is shy.*

Remark 5.6.4 In [PK2] (see proof of Theorem 4.1, Step 2) a partition $\Phi = \{A_\theta : \theta \in [0, 1]\}$ of the $\mathbf{R}^N$ into Haar ambivalent sets has been constructed such that for each $\theta_1, \theta_2 \in [0, 1]$ there exists an isometric (with respect to the Tychonoff metric which is invariant under translates) Borel measurable bijection $A_{(\theta_1,\theta_2)}$ of $\mathbf{R}^N$ such that $A_{(\theta_1,\theta_2)}(A_{\theta_1}) \Delta A_{\theta_2}$ is shy. In this context and regarding Lemma 5.6.1 it is natural to ask whether an arbitrary Polish nonlocally compact group with an invariant metric admits a similar partition into Haar ambivalent sets. Note that we have no information in this direction.

Theorem 5.6.1 *Let G be a Polish nonlocally compact group admitting an invariant metric. Then there exists a statistical structure* $\{(G, \mathbf{B}, \mu_\theta) : \theta \in \Theta\}$ *in G that has an objective consistent estimate of a parameter* θ *such that:*

(i) $\Theta \subseteq G$ *and* $card(\Theta) = 2^{\aleph_0}$.
(ii) μ_θ *is the restriction of the Dirac measure concentrated at the point* θ *to the Borel* σ*-algebra* $\mathbf{B}(G)$ *for each* $\theta \in \Theta$.

Proof By virtue of Lemma 5.6.1, there exists a closed set $F \subseteq G$ and a continuous function $\phi : F \to 2^N$ such that for any $x \in 2^N$ and any compact set $K \subseteq G$ there is $g \in G$ with $gK \subseteq \phi^{-1}(x)$. For $x \in 2^N \setminus \{(0, 0, \ldots)\}$ we put

$$X_x = \phi^{-1}(x). \tag{5.6.4}$$

We set $X_{(0,0,\ldots)} = \phi^{-1}((0, 0, \ldots)) \cup (G \setminus F)$. Thus we have a partition $\{X_x : x \in 2^N\}$ of G into Borel subsets such that each element of the partition is Borel measurable and a Haar ambivalent set. Let $\{\theta_x : x \in 2^N\}$ be any selector. We put $\Theta = \{\theta : \theta = \theta_x \text{ for some } x \in 2^N\}$ and denote by μ_θ the restriction of the Dirac measure concentrated at the point θ to the σ-algebra $\mathbf{B}(G)$. Thus we have constructed a statistical structure $\{(G, \mathbf{B}, \mu_\theta) : \theta \in \Theta\}$ in G. We put $T(g) = \theta$ for each $g \in X_\theta$. Now it is obvious that T is the objective consistent estimate of a parameter θ for the statistical structure $\{(G, \mathbf{B}, \mu_\theta) : \theta \in \Theta\}$ in G such that Conditions (i) and (ii) are fulfilled.

Theorem 5.6.2 *Let G be a Polish nonlocally compact group admitting an invariant metric. Let* μ *be a Borel probability measure whose carrier is a compact set* K_0*(i.e.,* $\mu(G \setminus K_0) = 0$*). Then there exists a statistical structure* $\{(G, \mathbf{B}, \mu_\theta) : \theta \in \Theta\}$ *in G that has an objective consistent estimate of a parameter* θ *such that*

(i) $\Theta \subseteq G$ *and* $card(\Theta) = 2^{\aleph_0}$.
(ii) μ_θ *is a* θ*-shift of the measure* μ *(i.e.,* $\mu_\theta(X) = \mu(\theta^{-1}X)$ *for* $X \in \mathbf{B}(G)$ *and* $\theta \in \Theta$*).*

Proof By virtue of Lemma 5.6.1, there exists a closed set $F \subseteq G$ and a continuous function $\phi : F \to 2^N$ such that for any $x \in 2^N$ and any compact set $K \subseteq G$ there is $g \in G$ with $gK \subseteq \phi^{-1}(x)$. For $x \in 2^N \setminus \{(0, 0, \dots)\}$ we put $X_x = \phi^{-1}(x)$. We set $X_{(0,0,\dots)} = \phi^{-1}((0, 0, \dots)) \cup (G \setminus F)$. Thus we have a partition $\{X_x : x \in 2^N\}$ of G into Borel subsets such that each element of the partition is Borel measurable, a Haar ambivalent set, and for any $x \in 2^N$ and any compact set $K \subseteq G$ there is $g \in G$ with $gK \subseteq X_x$. If we take a set K_0 under K, then for any $x \in 2^N$ there is $g(K_0, x) \in G$ with $g(K_0, x)K_0 \subseteq X_x$. We put $\Theta = \{\theta : \theta = g(K_0, x) \,\&\, x \in 2^N\}$. For each $\theta \in \Theta$ and $X \in \mathbf{B}(G)$, we put $\mu_\theta(X) = \mu(\theta^{-1}X)$. For $g \in X_x$ we put $T(g) = g(K_0, x)$. Let us show that $T : G \to \Theta$ is an objective consistent estimate of a parameter θ. Indeed, for each $\theta \in \Theta$ we have

$$\mu_\theta(T^{-1}(\theta)) = \mu_{g(K_0,x)}(T^{-1}(g(K_0, x))) = \mu_{g(K_0,x)}(X_x) = \mu(g(K_0, x)^{-1}X_x) \geq$$

$$\mu(g(K_0, x)^{-1}g(K_0, x)K_0) = \mu(K_0) = 1, \tag{5.6.5}$$

which means that $T : G \to \Theta$ is a consistent estimate of a parameter θ. On the other hand, for each $\theta = g(K_0, x) \in \Theta$ we have that a set $T^{-1}(\theta) = T^{-1}(g(K_0, x)) = X_x$ is Borel measurable and a Haar ambivalent set which together with formula (7.6.5) implies that $T : G \to \Theta$ is an objective consistent estimate of a parameter θ. Now it is obvious to check that for the statistical structure $\{(G, \mathbf{B}, \mu_\theta) : \theta \in \Theta\}$ Conditions (i) and (ii) are fulfilled.

The next theorem shows whether an objective consistent estimate can be constructed by virtue of some consistent estimates in a nonlocally compact Polish group admitting an invariant metric.

Theorem 5.6.3 *Assume G is a nonlocally compact Polish group admitting an invariant metric. Let card(Θ) $= 2^{\aleph_0}$ and $T : G \to \Theta$ be a consistent estimate of a parameter θ for the family of Borel probability measures $(\mu_\theta)_{\theta\in\Theta}$ such that there exists $\theta_0 \in \Theta$ for which $T^{-1}(\theta_0)$ is a prevalent set. Then there exists an objective consistent estimate of a parameter θ for the family $(\mu_\theta)_{\theta\in\Theta}$.*

Proof For $\theta \in \Theta$ we put $S_\theta = T^{-1}(\theta)$. Because S_{θ_0} is a prevalent set we deduce that

$$\cup_{\theta\in\Theta\setminus\{\theta_0\}} S_\theta = \mathbf{R}^N \setminus S_{\theta_0} \tag{5.6.6}$$

is shy in G.

By Lemma 5.2.3 we know that the measure μ_{θ_0} is concentrated on a union of a countable family of compact subsets $\{F_k^{(\theta_0)} : k \in N\}$. By Lemmas 5.2.1 and 5.6.2, we know that $\cup_{k\in N} F_k^{(\theta_0)}$ is shy in G.

We put $\tilde{S}_\theta = S_\theta$ for $\theta \in \Theta \setminus \{\theta_0\}$ and $\tilde{S}_{\theta_0} = \cup_{k\in N} F_k^{(\theta_0}$. Clearly, $S = \cup_{\theta\in\Theta} \tilde{S}_\theta$ is shy in G.

By virtue of Lemma 5.6.1, there exists a closed set $F \subseteq G$ and a continuous function $\phi : F \to 2^N$ such that for any $x \in 2^N$ and any compact set $K \subseteq G$ there is $g \in G$ with $gK \subseteq \phi^{-1}(x)$. Let $f : 2^N \to \Theta$ be any bijection. For $\theta \in \Theta$ we put

$$B_\theta = (\phi^{-1}(f^{-1}(\theta)) \setminus S) \cup \overline{S}_\theta. \tag{5.6.7}$$

Note that $(B_\theta)_{\theta \in \Theta}$ is a partition of G into Haar ambivalent sets. We put $T_1(g) = \theta$ for $g \in B_\theta (\theta \in \Theta)$. Because

$$\mu_\theta(T_1^{-1}(\theta)) = \mu_\theta(B_\theta) \geq \mu_\theta(\overline{S}_\theta) = 1 \tag{5.6.8}$$

for $\theta \in \Theta$, we claim that T_1 is a consistent estimate of a parameter θ for the family $(\mu_\theta)_{\theta \in \Theta}$. Because $T_1^{-1}(\theta)) = B_\theta$ is a Borel and Haar ambivalent set for each $\theta \in \Theta$ we end the proof of the theorem.

Example 5.6.4 Let F be a distribution function on $\mathbf{R}$ such that the integral $\int_{\mathbf{R}} x dF(x)$ exists and is equal to zero. Suppose that p is a Borel probability measure on $\mathbf{R}$ defined by F. For $\theta \in \Theta (:= \mathbf{R})$, let p_θ be a θ shift of the measure p(i.e., $p_\theta(X) = p(X - \theta)$ for $X \in \mathbf{B}(\mathbf{R})$). Setting, $G = \mathbf{R}^N$, for $\theta \in \Theta$ we put $\mu_\theta = p_\theta^N$, where p_θ^N denotes the infinite power of the measure p_θ. We set $T((x_k)_{k \in N}) = \lim_{n \to \infty} \frac{\sum_{k=1}^n x_k}{n}$, if $\lim_{n \to \infty} \frac{\sum_{k=1}^n x_k}{n}$ exists, is finite and differs from the zero, and $T((x_k)_{k \in N}) = 0$, otherwise. Note that $T : \mathbf{R}^N \to \Theta$ is a consistent estimate of a parameter θ for the family $(\mu_\theta)_{\theta \in \Theta}$ such that $T^{-1}(0)$ is a prevalent set. Indeed, by virtue of the strong law of large numbers, we know that

$$\mu_\theta\left(\left\{(x_k)_{k \in N} : \lim_{n \to \infty} \frac{\sum_{k=1}^n x_k}{n} = \theta\right\}\right) = 1 \tag{5.6.9}$$

for $\theta \in \Theta$.

Following [PK1] (Lemma 4.14, p. 60), a set S defined by

$$S = \left\{(x_k)_{k \in N} : \lim_{n \to \infty} \frac{\sum_{k=1}^n x_k}{n} \text{ exists and is finite}\right\}, \tag{5.6.10}$$

is a Borel shy set. This implies that $\mathbf{R}^N \setminus S$ is a prevalent set. Because $\mathbf{R}^N \setminus S \subseteq T^{-1}(0)$, we deduce that $T^{-1}(0)$ is a prevalent set. Because for the statistical structure $\{(\mathbf{R}^N, \mathbf{B}(\mathbf{R}^N), \mu_\theta) : \theta \in \Theta\}$ all conditions of the Theorem 7.6.3 are fulfilled, we claim that there exists an objective consistent estimate of a parameter θ for the family $(\mu_\theta)_{\theta \in \Theta}$.

Note that in Theorem 5.4.1 (see also [PK2], Theorem 3.1, p. 117) has been considered an example of a strong objective infinite sample consistent estimate of an unknown parameter for a certain statistical structure in the Polish nonlocally compact Abelian group $\mathbf{R}^N$. In the context of this example we state the following.

Problem 5.6.8 Let G be a Polish nonlocally compact group admitting an invariant metric. Does there exist a statistical structure $\{(G, \mathbf{B}(G), \mu_\theta) : \theta \in \Theta\}$ with $\mathrm{card}(\Theta) = 2^{\aleph_0}$ for which there exists a strong objective consistent estimate of a parameter θ?

5.7 Objective and Strong Objective Consistent Estimates of an Unknown Parameter in a Compact Polish Group $\{0,1\}^N$

Let $x_1, x_2, \ldots, x_k, \ldots$ be an infinite sample obtained by coin tosses. Then the statistical structure described in this experiment has the form:

$$\{(\{0,1\}^N, B(\{0,1\}^N), \mu_\theta^N) : \theta \in (0,1)\} \tag{5.7.1}$$

where $\mu_\theta(\{1\}) = \theta$ and $\mu_\theta(\{0\}) = 1-\theta$. By virtue of the strong law of large numbers we have

$$\mu_\theta^N\left(\left\{(x_k)_{k\in N} : (x_k)_{k\in N} \in \{0,1\}^N \ \&\ \lim_{n\to\infty}\frac{\sum_{k=1}^n x_k}{n} = \theta\right\}\right) = 1 \tag{5.7.2}$$

for each $\theta \in (0,1)$.

Note that for each $k \in N$, $G_k = \{0,1\}$ can be considered as a compact group with an addition group operation $\pmod 2$. Hence the space of all infinite samples $G := \{0,1\}^N$ can be presented as an infinite product of compact groups $\{G_k : k \in N\}$; that is, $G = \prod_{k\in N} G_k$. Note also that the group G admits an invariant metric ρ defined by $\rho((x_k)_{k\in N}, (y_k)_{k\in N}) = \sum_{k\in N} \frac{|x_k - y_k \pmod 2|}{2^{k+1}(1+|x_k-y_k \pmod 2|)}$ for $(x_k)_{k\in N}, (y_k)_{k\in N} \in G$. It is obvious that the measure λ_k on G_k defined by $\lambda_k(\{0\}) = \lambda_k(\{1\}) = 1/2$ is a probability Haar measure in G_k for each $k \in N$ and for the probability Haar measure λ in G the following equality $\lambda = \prod_{k\in N}\lambda_k$ holds true, and equivalently, $\lambda = \mu_{0,5}^N$.

By virtue of (5.7.2) we deduce that the set

$$A(0,5) = \left\{(x_k)_{k\in N} : (x_k)_{k\in N} \in \{0,1\}^N \ \&\ \lim_{n\to\infty}\frac{\sum_{k=1}^n x_k}{n} = 0,5\right\} \tag{5.7.3}$$

is prevalence. Because $A(\theta) \subset G \setminus A(0,5)$ for each $\theta \in (0;1) \setminus \{1/2\}$, where

$$A(\theta) = \left\{(x_k)_{k\in N} : (x_k)_{k\in N} \in \{0,1\}^N \ \&\ \lim_{n\to\infty}\frac{\sum_{k=1}^n x_k}{n} = \theta\right\}, \tag{5.7.4}$$

we deduce that they are all shy (equivalently, of Haar measure zero) sets. In terms of [HSY], this phenomenon can be expressed in the following form.

Theorem 5.7.1 *For almost every sequence* $(x_k)_{k\in N} \in \{0,1\}^N$ *its Cezaro means* $(\frac{\sum_{k=1}^n x_k}{n})_{n\in N}$ *converges to* $0,5$ *whenever* n *tends to* ∞.

By virtue of the strong law of large numbers, we get the following.

Theorem 5.7.2 *Fix* $\theta_0 \in (0,1)$. *For each* $(x_k)_{k\in N} \in G$, *we set* $T((x_k)_{k\in N}) = \lim_{n\to\infty}\frac{\sum_{k=1}^n x_k}{n}$ *if this limit exists and differs from* θ_0, *and* $T((x_k)_{k\in N}) = \theta_0$, *otherwise. Then* T *is a consistent estimate of an unknown parameter* θ *for the statistical structure* $\{(G, \mathbf{B}(G), \mu_\theta) : \theta \in \Theta\}$.

Remark 5.7.1 Following Definition 5.6.8, the estimate T is subjective because $T^{-1}(1/2)$ is a prevalent set. Unlike Theorem 5.6.3, there does not exist an objective consistent estimate of an unknown parameter θ for any statistical structure $\{(G, \mathbf{B}(G), \mu_\theta) : \theta \in \Theta\}$ for which $\mathrm{card}(\Theta) > \aleph_0$, where $\aleph_0$ denotes the cardinality of the set of all natural numbers. Indeed, assume the contrary and let T_1 be such an estimate. Then we get the partition $\{T_1^{-1}(\theta) : \theta \in \Theta\}$ of the compact group G into Haar ambivalent sets. Because each Haar ambivalent set is of positive λ measure, we get that the probability Haar measure λ does not satisfy the Suslin property provided that the cardinality of an arbitrary family of pairwise disjoint Borel measurable sets of positive λ measure in G is not more than countable.

Remark 5.7.2 Let us consider a mapping $F : G \to [0, 1]$ defined by $F((x_k)_{k\in N}) = \sum_{k\in N} \frac{x_k}{2^k}$ for $(x_k)_{k\in N} \in G$. This a Borel isomorphism between G and $[0, 1]$ such that the equality $\lambda(X) = \ell_1(F(X))$ holds true for each $X \in \mathbf{B}(G)$. By virtue of the latter relation, for each natural number m, there exists a partition $\{X_k : 1 \le k \le m\}$ of the group G into Haar ambivalent sets such that for each $1 \le i \le j \le m$ there is an isometric Borel measurable bijection $f_{(i,j)} : G \to G$ such that the set $f_{(i,j)}(X_i)\Delta X_j$ is shy, equivalently, of the λ measure zero.

By the scheme presented in the proof of Theorem 5.6.3, one can obtain the validity of the following assertions.

Theorem 5.7.3 *Let Θ_1 be a subset of the Θ with card$(\Theta) \ge 2$. Then there exists an objective consistent estimate of an unknown parameter θ for the statistical structure $\{(G, \mathbf{B}(G), \mu_\theta) : \theta \in \Theta_1\}$ if and only if card$(\Theta_1) \le \aleph_0$ and $1/2 \notin \Theta_1$.*

Theorem 5.7.4 *Let Θ_2 be a subset of the Θ card$(\Theta) \ge 2$. Then there exists a strong objective consistent estimate of an unknown parameter θ for the statistical structure $\{(G, \mathbf{B}(G), \mu_\theta) : \theta \in \Theta_2\}$ if and only if card$(\Theta_2) < \aleph_0$ and $1/2 \notin \Theta_2$.*

Remark 5.7.3 Main results of Chap. 5 were obtained in [KKP].

5.8 Conclusion

Let NHST be associated with a statistical structure $(G, \mathcal{B}(G), \mu_\theta)_{\theta\in\Theta}$ and a consistent estimation $T : G \to \Theta$, when G is a Polish group equipped with an invariant metric. A certain claim in that each consistent estimation T must pass the certification exam on objectivity before its practical application can be given as follows. First, note that if there is a parameter $\theta_0 \in \Theta$ for which $T^{-1}(\theta_0)$ is a shy set in G, then the acceptance region for the parameter θ_0 under null hypothesis $H_0 : \theta = \theta_0$ will be small and correspondingly, for almost every (in the sense of [HSY]) sample we reject the null hypothesis $H_0 : \theta = \theta_0$ losing an objectivity of the hypothesis testing. Just here it is pertinent to mention the quote from [N]: "... it is usually nonsensical to perform an experiment with the sole aim of rejecting the null hypothesis." Second, note that if there is a parameter $\theta_1 \in \Theta$ for which $T^{-1}(\theta_1)$ is a prevalent set in G then the acceptance region for parameter θ_1 under null hypothesis $H_0 : \theta = \theta_1$ will be large and correspondingly, for almost every (in the sense of [HSY]) sample we accept null hypothesis $H_0 : \theta = \theta_1$ again losing an objectivity of the hypothesis testing. In that case, I conjecture that *"It is usually nonsensical also to perform an experiment with the sole aim of accepting the null hypothesis"*, and really, it is difficult-won't agree with me. Not losing an objectivity of the hypothesis testing, for each $\theta \in \Theta$ a set $T^{-1}(\theta)$ must be neither shy nor a prevalent set in G which [BBE] called a Haar ambivalent set in G. The latter observation directly leads us to the notion of the objectivity of the consistent estimation T and NHST, respectively.

We briefly discuss what we recommend to statisticians from the point of view of practical applications of NHST at this stage:

- Recommendation 5.8.1. If G is locally compact and $\text{card}(\Theta) > \aleph_0$, then do not try to construct an objective NHST because it does not exist. (This claim is similar to the proof of Remark 5.7.1.)
- Recommendation 5.8.2. If G is nonlocally compact then the consistent estimation T must pass an exam on the objectivity provided that we must choose the validity of the condition

$$(\forall\theta)(\theta \in \Theta \rightarrow T^{-1}(\theta) \text{ a Haar ambivalent set in } G). \tag{5.8.1}$$

- Recommendation 5.8.3. If G is nonlocally compact and the consistent estimation T satisfies Condition (5.8.1) then NHST is objective and we can use it for our practical purposes.
- Recommendation 5.8.4. If G is nonlocally compact and the consistent estimation T does not satisfy Condition (5.8.1) but satisfies the condition

$$(\exists\theta_0)(\theta_0 \in \Theta \rightarrow T^{-1}(\theta_0) \text{ is prevalent set}), \tag{5.8.2}$$

then we must construct its modification $\overline{T}$ satisfying Condition (5.8.1; cf. Theorem 5.6.3), and only after this can NHST (with the objective estimation $\overline{T}$) be used for our practical purposes.

References

[BBE] Balka, R., Buczolich, Z., Elekes, M.: Topological hausdorff dimension and level sets of generic continuous functions on fractals. Chaos Solitons Fractals **45**(12), 1579–1589 (2012)

[Chr1] Christensen, J.R.: Measure theoretic zero sets in infinite dimensional spaces and applications to differentiability of Lipschitz mappings. Publ. Dep. Math. **10**(2), 29–39 (1973)

[Coh] Cohen, J.: The Earth is round ($p < .05$). Am. Psychol. **49**(12), 997–1003 (1994)

[D] Dougherty, R.: Examples of non-shy sets. Fund. Math. **144**, 73–88 (1994)

[HSY] Hunt, B.R., Sauer, T., Yorke, J.A.: Prevalence: a translation-invariant "almost every" on infinite-dimensional spaces. Bull. Am. Math. Soc. **27**(2), 217–238 (1992)

[IS] Ibramkhallilov, I.S., Skorokhod, A.V.: On Well-off Estimates of Parameters of Stochastic Processes (in Russian). Naukova Dumka, Kiev (1980)

[K] Kaisan, C.: High accuracy calculation, Cauchy distribution (percentile). http://keisan.casio.com/has10/SpecExec.cgi

[Kh] Kharazishvili, A.B.: Topologicheskie aspekty teorii mery (Russian). [Topological aspects of measure theory]. Naukova Dumka, Kiev (1984)

[KKP] Kintsurashvili, M., Kiria, T., Pantsulaia, G.: On objective and strong objective consistent estimates of unknown parameters for statistical structures in a Polish group admitting an invariant metric. J. Stat. Adv. Theory Appl. **13**(2), 179–233 (2015)

[KN] Kuipers, L., Niederreiter, H.: Uniform Distribution of Sequences. Wiley, New York (1974)

[MS] Mycielski, J., Swierczkowski, S.: On the Lebesgue measurability and the axiom of determinateness. Fund. Math. **54**, 67–71 (1964)

[M] Mycielski, J.: Some unsolved problems on the prevalence of ergodicity, instability, and algebraic independence. Ulam Quart. 1(3). 30 ff. approx. p. 8 (1992) (electronic only)

[N] Nadaraya, E.: On non-parametric estimates of density functions and regression curves. Theor. Prob. Appl. **10**, 186–190 (1965)

[N] Nunnally, J.: The place of statistics in psychology. Educ. Psychol. Measur. **20**(4), 641–650 (1960)

[P1] Pantsulaia, G.R.: On separation properties for families of probability measures. Georgian Math. J. **10**(2), 335–342 (2003)

[P2] Pantsulaia, G.R.: Invariant and quasiinvariant measures in infinite-dimensional topological vector spaces. Nova Science Publishers Inc., New York (2007)

[P3] Pantsulaia, G.: On a certain partition of the non-locally compact abelian polish group R^N. Proc. A. Razmadze Math. Inst. **149**, 75–86 (2009)

[PK1] Pantsulaia, G., Kintsurashvili, M.: Why is null hypothesis rejected for "almost every" infinite sample by some hypothesis testing of maximal reliability? J. Stat. Adv. Theory Appl. **11**(1), 45–70 (2014). www.scientificadvances.co.in

[PK2] Pantsulaia, G., Kintsurashvili, M.: An effective construction of the strong objective infinite sample well-founded estimate. Proc. A. Razmadze Math. Ins. **166**, 113–119 (2014)

[PK3] Pantsulaia, G., Kintsurashvili, M.: An objective infinite sample well-founded estimates of a useful signal in the linear one-dimensional stochastic model Rep. Enlarged Sess. Semin. I. Vekua. Appl. Math. **28**, 90–93 (2014)

[R] Rosenblatt, M.: Remarks on some nonparametric estimates of a density function. Ann. Math. Stat. **27**, 832–837 (1956)

[Sc] Schuster, E.F.: Estimation of a probability density function and its derivatives. Ann. Math. Stat. **40**, 1187–1195 (1969)

[Sh] Shiryaev, A.N.: Problems in Probability. Problem books in mathematics. Springer, New York (2012)

[Solovay] Solovay, R.M.: A model of set theory in which every set of reals is Lebesgue measurable. Ann. Math. **92**, 1–56 (1970)

[Sol] Solecki, S.: On haar null sets. Fund. Math. **149**(3), 205–210 (1996)

[ZPS] Zerakidze, Z., Pantsulaia, G., Saatashvili, G.: On the separation problem for a family of Borel and Baire G-powers of shift-measures on R. Ukr. Math. J. **65**(4), 470–485 (2013)

Chapter 6
Why Null Hypothesis Is Rejected for Almost Every Infinite Sample by the Hypothesis Testing of a Maximal Reliability

6.1 Introduction

Criticism of statistical hypothesis testing has been considered in [Ch, LMS, K, MZ, MH, O], citing 300–400 primary references. Many of the philosophical criticisms of hypothesis testing contain the general point of view that the theory of mathematical statistics and results of testing are inconsistent in many situations. Various different reasonable statistical methods in many expensive experiments led to inconsistent decisions, causing great alarm among mathematicians and statisticians. It would be ridiculous and practically impossible to explain all the paradoxes that underlie the existence of a big gap between theory and practice. Here we do not consider in detail all the issues that summarize such a criticism. We focus our attention on a certain confusion described in the works of Jum Nunnally [N] and Jacob Cohen [Coh]. As quoted earlier, in [Coh] Jacob Cohen said, "...Don't look for a magic alternative to NHST [null hypothesis significance testing] ... It does not exist." In [N] Jum Nunnally conjectured,

> If the decisions are based on convention they are termed arbitrary or mindless while those not so based may be termed subjective. To minimize type *II* errors, large samples are recommended. In psychology practically all null hypotheses are claimed to be false for sufficiently large samples so ... it is usually nonsensical to perform an experiment with the sole aim of rejecting the null hypothesis.

A question naturally arises here as to whether concepts of the theory of statistical decisions for infinite samples can be introduced and whether Jacob Cohen's and Jum Nunnally's above-mentioned conjectures are valid. To confirm the validity of their predictions, for a linear one-dimensional stochastic system, we consider a certain hypothesis testing for infinite samples such that the sum of errors of I and II types is equal to zero. (We call such tests, tests of a maximal reliability.) Furthermore, we explain why a null hypothesis is claimed to be false for almost every [HSY] infinite sample.

First, note that in many cases even information advising us that some phenomenon happens with probability 1 can be quite poor can be considered an essential reason

G. Pantsulaia, *Applications of Measure Theory to Statistics*,
DOI 10.1007/978-3-319-45578-5_6

for all inconsistent statistical decisions. Indeed, let X be an infinite-dimensional topological vector space. Let P be any sentence formulated for elements in X and let μ be any probability Borel measure on X. Let us discuss what information the following sentence yields.

"*μ-almost every element of X satisfies the property P.*"

If X is separable then an arbitrary nonzero σ-finite Borel measure defined on X is concentrated on the union of countable compact subsets $(F_k)_{k \in \mathbb{N}}$ in B (cf. [Kh]) and for arbitrary $k \in \mathbb{N}$ there exists a vector $v_k \in X$ which spans a line L_k such that every translation of L_k meets F_k in at most one point (cf.[70], p.225, Fact 8). In such a way, the support of μ may be regarded as the union of a countable family of "surfaces". Therefore the information described by the above-mentioned sentence, in general, may be very poor. For this reason to study the behavior of various general systems defined in infinite-dimensional separable topological vector spaces in terms of any partial σ-finite Borel measure (e.g., Gaussian measure that is concentrated on a poor set) is not recommended and needs to extend the measure-theoretic terms "measure zero" and "almost every." This phenomenon was first noted by J. Christensen (1973) [Ch1] and more lately by B.R. Hunt, T. Sauer, J.A. Yorke (1992) [HSY], J. Mycielski (1992) [M], R. Dougherty (1994) [D], and other well-known mathematicians. As with the concept of "Lebesgue almost every" on finite-dimensional spaces, their notion of "prevalence" is translation-invariant. Instead of using a specific measure on the entire space, they have defined prevalence in terms of the class of all probability measures with compact support. Prevalence is a more appropriate condition than the topological concepts of "open and dense" or "generic" when one desires a probabilistic result on the likelihood of a given property on a function space.

The purpose of the present chapter is an application of the approach of "almost every" in studying structures of domains of some infinite sample statistics and in explaining why the null hypothesis is rejected for "almost every" infinite sample by the hypothesis testing of a maximal reliability.

The rest of this chapter is the following.

In Sect. 6.2 we give auxiliary notions and facts from functional analysis and measure theory.

In Sect. 6.3 we consider consistent estimators of a useful signal in the linear one-dimensional stochastic model when the expectation of the transformed signal is not defined.

In Sect. 6.4 we consider an example of hypothesis testing of high reliability for a linear one-dimensional stochastic model and explain why the null hypothesis is rejected for "almost every" infinite sample. Also, we consider some well-known infinite sample statistics and study structures of their domains in terms of "prevalence."

6.2 Auxiliary Notions and Facts from Functional Analysis and Measure Theory

Let $\mathbb{V}$ be a complete metric linear space, by which we mean a vector space (real or complex) with a complete metric for which addition and scalar multiplication are continuous. When we speak of a measure on $\mathbb{V}$ we always mean a nonnegative measure that is defined on the Borel sets of $\mathbb{V}$ and is not identically zero. We write $S + v$ for the translate of a set $S \subseteq \mathbb{V}$ by a vector $v \in \mathbb{V}$.

Definition 6.2.1 ([HSY], *Definition 1, p. 221*) A measure μ is said to be transverse to a Borel set $S \subset \mathbb{V}$ if the following two conditions hold.

(i) there exists a compact set $U \subset \mathbb{V}$ for which $0 < \mu(U) < 1$;
(ii) $\mu(S + v) = 0$ for every $v \in \mathbb{V}$.

Definition 6.2.2 ([HSY], *Definition 2, p. 222*) A Borel set $S \subset \mathbb{V}$ is called shy if there exists a measure transverse to S. More generally, a subset of $\mathbb{V}$ is called shy if it is contained in a shy Borel set. The complement of a shy set is called a prevalent set.

Definition 6.2.3 ([HSY], *p. 226*) We say "almost every" element of $\mathbb{V}$ satisfies a given property, if the subset of $\mathbb{V}$ on which the property holds is prevalent.

Lemma 6.2.1 ([HSY], Fact 3″, p. 223) *The union of a countable collection of shy sets is shy.*

Lemma 6.2.2 ([HSY], Fact 8, p. 224) *If $\mathbb{V}$ is infinite-dimensional, all compact subsets of $\mathbb{V}$ are shy.*

Definition 6.2.4 ([HSY], *Definition 6, p. 225*) We call a finite-dimensional subspace $P \subset \mathbb{V}$ a probe for a set $T \subset \mathbb{V}$ if a Lebesgue measure supported on P is transverse to a Borel set that contains the complement of T.

Remark 6.2.1 Note that a sufficient (but not necessary) condition for T to be prevalent is for it to have a probe.

One can consult [HSY] in order to see whether the validity of the following assertions can be obtained by constructing the appropriate probes.

Example 6.2.1 ([HSY], *Proposition 1, p. 226*) "Almost every" function $f : [0; 1] \to \mathbb{R}$ in L_1 satisfies $\int_0^1 f(x)dx \neq 0$.

Example 6.2.2 ([HSY], *Proposition 2, p. 226*) For $1 < p \leq \infty$ "almost every" sequence $(a_i)_{i \in \mathbb{N}}$ in ℓ^p has the property that $\sum_{i=1}^{\infty} a_i$ diverges.

Example 6.2.3 ([HSY], *Proposition 4, p. 226*) "Almost every" continuous function $f : [0, 1] \to \mathbb{R}$ is nowhere differentiable.

Lemma 6.2.3 ([Kh], Lemma 2, p. 58) *Let μ be a Borel probability measure defined in complete separable metric space $\mathbb{V}$. Then there exists a countable family of compact sets $(F_k)_{k \in \mathbb{N}}$ in $\mathbb{V}$ such that*

$$\mu(\mathbb{V} \setminus \cup_{k \in \mathbb{N}} F_k) = 0.$$

6.3 Consistent Estimators of a Useful Signal in a One-Dimensional Linear Stochastic Model When the Expectation of the Transformed Signal Is Not Defined

Suppose that Θ is a vector subspace of the infinite-dimensional topological vector space of all real-valued sequences $\mathbb{R}^{\mathbb{N}}$ equipped with the product topology.

In information transmission theory we consider the linear one-dimensional stochastic system

$$(\xi_k)_{k\in\mathbb{N}} = (\theta_k)_{k\in\mathbb{N}} + (\Delta_k)_{k\in\mathbb{N}}, \tag{6.3.1}$$

where $(\theta_k)_{k\in\mathbb{N}} \in \Theta$ is a sequence of useful signals, $(\Delta_k)_{k\in\mathbb{N}}$ is a sequence of independent identically distributed random variables (the so-called generalized "white noise") defined on some probability space $(\Omega, \mathbf{F}, P)$, and $(\xi_k)_{k\in\mathbb{N}}$ is a sequence of transformed signals. Let μ be a Borel probability measure on $\mathbb{R}$ defined by a random variable Δ_1. Then the $\mathbb{N}$ power of the measure μ denoted by $\mu^{\mathbb{N}}$ coincides with the Borel probability measure on $\mathbb{R}^{\mathbb{N}}$ defined by the generalized "white noise"; that is,

$$(\forall X)(X \in \mathbf{B}(\mathbb{R}^{\mathbb{N}}) \rightarrow \mu^{\mathbb{N}}(X) = P(\{\omega : \omega \in \Omega \ \& \ (\Delta_k(\omega))_{k\in\mathbb{N}} \in X\})), \tag{6.3.2}$$

where $\mathbf{B}(\mathbb{R}^{\mathbb{N}})$ is the Borel σ-algebra of subsets of $\mathbb{R}^{\mathbb{N}}$.

In information transmission theory, the general decision is that the Borel probability measure λ, defined by the sequence of transformed signals $(\xi_k)_{k\in\mathbb{N}}$ coincides with $\left(\mu^{\mathbb{N}}\right)_{\theta_0}$ for some $\theta_0 \in \Theta$ provided that

$$(\exists\theta_0)(\theta_0 \in \Theta \rightarrow (\forall X)(X \in \mathbf{B}(\mathbb{R}^{\mathbb{N}}) \rightarrow \lambda(X) = \left(\mu^{\mathbb{N}}\right)_{\theta_0}(X))), \tag{6.3.3}$$

where $\left(\mu^{\mathbb{N}}\right)_{\theta_0}(X) = \mu^{N}(X - \theta_0)$ for $X \in \mathbf{B}(\mathbb{R}^{\mathbb{N}})$.

Here we consider a particular case of the above model when a vector space of useful signals Θ has the form

$$\Theta = \{(\theta, \theta, \ldots) : \theta \in \mathbb{R}\}. \tag{6.3.4}$$

For $\theta \in \mathbb{R}$, a measure $\mu_\theta^{\mathbb{N}}$, defined by

$$\mu_\theta^{N} = \mu_\theta \times \mu_\theta \times \cdots,$$

where μ_θ is a θ-shift of μ (i.e., $\mu_\theta(X) = \mu(X - \theta)$ for $X \in \mathbf{B}(\mathbb{R})$), is called the $\mathbb{N}$ power of the θ-shift of μ on $\mathbb{R}$. It is obvious that $\mu_\theta^{\mathbb{N}} = \left(\mu^{\mathbb{N}}\right)_{(\theta,\theta,\ldots)}$.

As usual, the following main statistical decision, a triplet $(\mathbb{R}^{\mathbb{N}}, \mathbf{B}(\mathbb{R}^{\mathbb{N}}), \mu_\theta^{\mathbb{N}})_{\theta\in\mathbb{R}}$ is called a statistical structure describing the linear one-dimensional stochastic system (6.3.1).

Definition 6.3.1 A Borel measurable function $T_n : \mathbb{R}^n \to \mathbb{R}$ $(n \in \mathbb{N})$ is called a consistent estimator of a parameter θ for the family $(\mu_\theta^{\mathbb{N}})_{\theta\in\mathbb{R}}$ if the condition

$$\mu_\theta^{\mathbb{N}}(\{(x_k)_{k\in\mathbb{N}} : (x_k)_{k\in\mathbb{N}} \in \mathbb{R}^{\mathbb{N}} \ \& \ \lim_{n\to\infty} T_n(x_1,\dots,x_n) = \theta\}) = 1 \qquad (6.3.5)$$

holds for each $\theta \in \mathbb{R}$.

Theorem 6.3.1 *Let F be a strictly increasing continuous distribution function on $\mathbb{R}$ and μ be a Borel probability measure on $\mathbb{R}$ defined by F. For $\theta \in \mathbb{R}$, we set $F_\theta(x) = F(x-\theta)(x \in \mathbb{R})$ and denote by μ_θ the Borel probability measure on $\mathbb{R}$ defined by F_θ (obviously, this is an equivalent definition of the θ-shift of μ). Then a function $T_n : \mathbb{R}^n \to \mathbb{R}$, defined by*

$$T_n(x_1,\dots,x_n) = -F^{-1}(n^{-1}\#(\{x_1,\dots,x_n\} \cap (-\infty; 0])) \qquad (6.3.6)$$

for $(x_1,\dots,x_n) \in \mathbb{R}^n$ $(n \in \mathbb{N})$, is a consistent estimator of a parameter θ for the family $(\mu_\theta^{\mathbb{N}})_{\theta\in\mathbb{R}}$.

Definition 6.3.2 Following [IS], the family $(\mu_\theta^{\mathbb{N}})_{\theta\in\mathbb{R}}$ is called strongly separated in the usual sense if there exists a family $(Z_\theta)_{\theta\in\mathbb{R}}$ of Borel subsets of $\mathbb{R}^{\mathbb{N}}$ such that

(i) $\mu_\theta^{\mathbb{N}}(Z_\theta) = 1$ for $\theta \in \mathbb{R}$.
(ii) $Z_{\theta_1} \cap Z_{\theta_2} = \emptyset$ for all different parameters θ_1 and θ_2 from $\mathbb{R}$.
(iii) $\cup_{\theta\in\mathbb{R}} Z_\theta = \mathbb{R}^{\mathbb{N}}$.

Definition 6.3.3 Following [IS], a Borel measurable function $T : \mathbb{R}^{\mathbb{N}} \to \mathbb{R}$ is called an infinite sample consistent estimator of a parameter θ for the family $(\mu_\theta^{\mathbb{N}})_{\theta\in\mathbb{R}}$ if the condition

$$(\forall\theta)(\theta \in \mathbb{R} \to \mu_\theta^{\mathbb{N}}(\{(x_k)_{k\in\mathbb{N}} : (x_k)_{k\in\mathbb{N}} \in \mathbb{R}^{\mathbb{N}} \ \& \ T((x_k)_{k\in\mathbb{N}}) = \theta\}) = 1) \qquad (6.3.7)$$

is fulfilled.

Remark 6.3.1 The existence of an infinite sample consistent estimator of a parameter θ for the family $(\mu_\theta^{\mathbb{N}})_{\theta\in\mathbb{R}}$ implies that the family $(\mu_\theta^{\mathbb{N}})_{\theta\in\mathbb{R}}$ is strongly separated in a usual sense. Indeed, if we set $Z_\theta = \{(x_k)_{k\in\mathbb{N}} : (x_k)_{k\in\mathbb{N}} \in \mathbb{R}^{\mathbb{N}} \ \& \ T((x_k)_{k\in\mathbb{N}}) = \theta\}$ for $\theta \in \mathbb{R}$, then all the conditions of Definition 6.3.3 will be satisfied.

By using the strong law of large numbers one can easily obtain the validity of the following assertion.

Lemma 6.3.1 *Let F be a strictly increasing continuous distribution function on $\mathbb{R}$ and μ be the Borel probability measure on $\mathbb{R}$ defined by F. Suppose that the first-order absolute moment of μ is finite and the first-order moment of μ is equal to zero. For $\theta \in \mathbb{R}$, we set $F_\theta(x) = F(x-\theta)(x \in \mathbb{R})$ and denote by μ_θ the Borel probability measure on $\mathbb{R}$ defined by F_θ. Then the estimators $\overline{\lim}\widetilde{T}_n := \inf_n \sup_{m\geq n} \widetilde{T}_m$ and*

$\underline{\lim}\widetilde{T}_n := \sup_n \inf_{m \geq n} \widetilde{T_m}$ *are infinite sample consistent estimators of a parameter* θ *for the family* $(\mu_\theta^{\mathbb{N}})_{\theta \in \mathbb{R}}$*, where* $\widetilde{T}_n : \mathbb{R}^{\mathbb{N}} \to \mathbb{R}$ *is defined by*

$$(\forall (x_k)_{k\in\mathbb{N}})\left((x_k)_{k\in\mathbb{N}} \in \mathbb{R}^{\mathbb{N}} \to \widetilde{T}_n((x_k)_{k\in\mathbb{N}}) = n^{-1}\sum_{k=1}^{n} x_k\right). \tag{6.3.8}$$

Lemma 6.3.2 ([ZPS], Theorem 4.2, p. 483) *Let F be a strictly increasing continuous distribution function on* $\mathbb{R}$ *and* μ *be the Borel probability measure on* $\mathbb{R}$ *defined by F. For* $\theta \in \mathbb{R}$*, we set* $F_\theta(x) = F(x-\theta)(x \in \mathbb{R})$ *and denote by* μ_θ *the Borel probability measure on* $\mathbb{R}$ *defined by* F_θ*. Then the estimators* $\overline{\lim}\widetilde{T}_n := \inf_n \sup_{m\geq n} \widetilde{T_m}$ *and* $\underline{\lim}\widetilde{T}_n := \sup_n \inf_{m\geq n} \widetilde{T_m}$ *are infinite sample consistent estimators of a parameter* θ *for the family* $(\mu_\theta^{\mathbb{N}})_{\theta\in\mathbb{R}}$*, where* $\widetilde{T}_n : \mathbb{R}^{\mathbb{N}} \to \mathbb{R}$ *is defined by*

$$(\forall (x_k)_{k\in\mathbb{N}})((x_k)_{k\in\mathbb{N}} \in \mathbb{R}^{\mathbb{N}} \to \widetilde{T}_n((x_k)_{k\in\mathbb{N}}) = -F^{-1}(n^{-1}\#(\{x_1, \ldots, x_n\} \cap (-\infty; 0]))). \tag{6.3.9}$$

Remark 6.3.2 By Remark 6.3.1 and Lemmas 6.3.1 and 6.3.2, we deduce that the families of powers of shift measures described in corresponding lemmas are strongly separated in the usual sense.

6.4 Some Statistical Tests and Their Criticism from the Point of View of Functional Analysis

Let us recall some notions of the theory of statistical decisions.

Let $(\mathbb{R}^{\mathbb{N}}, \mathbf{B}(\mathbb{R}^{\mathbb{N}}), \mu_\theta^{\mathbb{N}})_{\theta\in\mathbb{R}}$ be a statistical structure described by the linear one-dimensional stochastic system (6.3.1).

Definition 6.4.1 Let a null hypothesis H_0 be defined by $H_0 : \theta = \theta_0$, where $\theta \in \mathbb{R}$.

A triplet (T, U_0, U_1), where

(i) $T : \mathbb{R}^{\mathbb{N}} \to \mathbb{R}$ is a statistic (equivalently, Borel measurable function);
(ii) $U_0 \cup U_1 = \mathbb{R}^{\mathbb{N}}$, $U_0 \cap U_1 = \emptyset$ and $U_0 \in \mathbf{B}(\mathbb{R}^{\mathbb{N}})$

is called a statistical test (or criterion) for acceptance of null hypothesis H_0(or equivalently, HT(hypothesis testing)).

For (an infinite) sample $x \in \mathbb{R}^{\mathbb{N}}$, we accept null hypothesis H_0 if $T(x) \in U_0$ and reject it otherwise.

T is called a statistic of the criterion (T, U_0, U_1).

U_0 is called the region of acceptance for null hypothesis H_0.

U_1 is called the region of rejection (equivalently, critical region) for null hypothesis H_0.

Definition 6.4.2 A decision obtained by the criterion (T, U_0, U_1) is called an, if null hypothesis H_0 has been rejected whenever null hypothesis H_0 was true.

Definition 6.4.3 A decision obtained by the criterion (T, U_0, U_1) is called an error of type *II*, if null hypothesis H_0 has been accepted whenever null hypothesis H_0 was false.

Definition 6.4.4 The value

$$\mu_\theta^{\mathbb{N}}(\{x : T(x) \in U_1 | H_0\}) = \alpha \tag{6.4.1}$$

is called the size (equivalently, significance level) of the test T.

Definition 6.4.5 The value

$$\mu_\theta^{\mathbb{N}}(\{x : T(x) \in U_0 | H_1\}) = \beta \tag{6.4.2}$$

is called the power of the test T.

In many cases it is not possible to reduce values α and β simultaneously. For this reason we fix the probability α of the error of type *I* and consider such critical regions U_1 for which the following condition

$$\mu_\theta^{\mathbb{N}}(\{x : T(x) \in U_1 | H_0\}) \leq \alpha \tag{6.4.3}$$

holds. Furthermore, between such critical regions we choose a region U_1^* for which the error of type *II* is maximal.

We show that for a model (6.3.1) there exists a "good" HT (equivalently, HT of a maximal reliability) for which $\alpha + \beta = 0$. Furthermore, in terms "almost every" introduced in Definition 6.2.3, we try to explain why an application of such a "good" HT leads us to confusion.

Example 6.4.1 Let us consider the linear one-dimensional stochastic system (6.3.1) for which F is a linear standard Gaussian (or Cauchy) distribution function on $\mathbb{R}$.

For $\theta \in \mathbb{R}$ we put

$$D_\theta = \{(x_k)_{k\in\mathbb{N}} : (x_k)_{k\in\mathbb{N}} \in \mathbb{R}^{\mathbb{N}} \ \& \ \overline{\lim}\widetilde{T}_n((x_k)_{k\in\mathbb{N}}) = \theta\} \tag{6.4.4}$$

where the estimator $\overline{\lim}\widetilde{T}_n$ comes from Lemma 6.3.1 (or Lemma 6.3.2). By Lemma 6.3.1 (or Lemma 6.3.2) we know that

$$\mu_\theta^{\mathbb{N}}(D_\theta) = 1. \tag{6.4.5}$$

On the other hand, by Lemma 6.2.3 we know that for each $\theta \in \Theta$, there exists a countable family of compact sets $(F_k^{(\theta)})_{k\in\mathbb{N}}$ such that

$$\mu_\theta^{\mathbb{N}}(\mathbb{R}^{\mathbb{N}} \setminus \cup_{k\in\mathbb{N}} F_k^{(\theta)}) = 0. \tag{6.4.6}$$

Finally, for $\theta \in \Theta$ we put

$$C_\theta = D_\theta \cap \cup_{k \in \mathbb{N}} F_k^{(\theta)}. \tag{6.4.7}$$

It is obvious that $(C_\theta)_{\theta \in \Theta}$ is a family of pairwise disjoint F_σ sets such that

$$\mu_\theta^{\mathbb{N}}(C_\theta) = 1. \tag{6.4.8}$$

Regarding Example 6.4.1, we consider the following two statistical tests.

Test 6.4.1 (The decision rule for null hypothesis $H_0 : \theta = \theta_0$)
Null Hypothesis: $H_0 : \theta = \theta_0$
Alternative Hypothesis: $H_1 : \theta \neq \theta_0$
Test Statistic: $T((x_k)_{k \in \mathbb{N}}) = \overline{\lim} \widetilde{T}_n$.
Alternative Critical Region: $U_1 = \mathbb{R}^{\mathbb{N}} \setminus C_{\theta_0}$

Test 6.4.2 (The decision rule for a countable competing hypothesis $\{H_i : \theta = \theta_i : i \in \mathbb{N}\}$)
ith Hypothesis: $H_i : \theta = \theta_i$
Test Statistic: $T((x_k)_{k \in \mathbb{N}}) = \overline{\lim} \widetilde{T}_n$.
Acceptance Region For H_i: $U_i = C_{\theta_i}$
Alternative Critical Region: $V = \mathbb{R}^{\mathbb{N}} \setminus \cup_{i \in \mathbb{N}} U_i$

Because the family of probability measures $(\mu_\theta^{\mathbb{N}})_{\theta \in \mathbb{R}}$ is strongly separated (see Remark 6.3.2), by the general assumption (6.3.3) we deduce that the sum of errors of *I* and *II* types for Test 6.4.1 is equal to zero. However, the following result is valid.

Theorem 6.4.1 *For "almost every" infinite sample the null hypothesis is rejected by Test 6.4.1.*

Proof We have to show that an alternative critical region U_1 is prevalent because it coincides with a set of all samples for which the null hypothesis is rejected by Test 6.4.1. Because the set C_{θ_0} is covered by the union of the countable family of compact sets $(F_k^{(\theta_0)})_{k \in \mathbb{N}}$, by Lemmas 6.2.1 and 6.2.2 we deduce that a Borel set C_{θ_0} as a subset of the Borel shy set $\cup_{k \in \mathbb{N}} F_k^{(\theta_0)}$(see Definition 6.2.2) is also shy. The latter relation implies that U_1 is prevalent because it is a complement of the shy set C_{θ_0}. This ends the proof of the theorem.

Because the family of probability measures $(\mu_\theta^{\mathbb{N}})_{\theta \in \mathbb{R}}$ is strongly separated (see Remark 6.3.2), by the general assumption (6.3.3) we deduce that the sum of errors of all types for Test 6.4.2 is equal to zero. However, the following result is valid.

Theorem 6.4.2 *For "almost every" infinite sample each hypothesis $H_i (i \in \mathbb{N})$ is rejected by Test 6.4.2.*

Proof We have to show that an alternative critical region V is prevalent because it coincides with a set of all samples for which $H_i(i \in \mathbb{N})$ is rejected by Test 6.4.2. Because for $i \in \mathbb{N}$ the set C_{θ_i} is covered by the union of the countable family of compact sets $(F_k^{(\theta_i)})_{k\in\mathbb{N}}$, by Lemmas 6.2.1 and 6.2.2 we deduce that a set C_{θ_i} as a subset of the Borel shy set $\cup_{k\in\mathbb{N}}F_k^{(\theta_i)}$ (see Definitions 6.2.1 and 6.2.2) is also shy. By Lemmas 6.2.1 and 6.2.2 we know that $\cup_{i\in\mathbb{N}}U_i = \cup_{i\in\mathbb{N}}C_{\theta_i}$ is a shy set which implies that V as a complement of the shy set $\cup_{i\in\mathbb{N}}U_i$ is prevalent. This ends the proof of the theorem.

Remark 6.4.1 Note that Theorems 6.4.1 and 6.4.2 well explain the meaning of Jum Nunnally's conjecture [N] which asserted that … *in psychology practically all null hypotheses are claimed to be false for sufficiently large samples* ….

Theorem 6.4.3 *Let $T : \mathbb{R}^{\mathbb{N}} \to \mathbb{R}$ be an infinite sample average defined by*

$$T((x_k)_{k\in\mathbb{N}}) = \lim_{n\to\infty} \frac{\sum_{k=1}^{n} x_k}{n}. \tag{6.4.9}$$

Then for "almost every" infinite sample the statistic T does not exist.

Proof Let S be defined by

$$S = \{(x_k)_{k\in\mathbb{N}} : (x_k)_{k\in\mathbb{N}} \in \mathbb{R}^{\mathbb{N}} \ \&\ \text{there exists a finite limit} \lim_{n\to\infty} \frac{\sum_{k=1}^{n} x_k}{n}\}. \tag{6.4.10}$$

It is obvious that S is a vector subspace of $\mathbb{R}^{\mathbb{N}}$. Indeed, if $(x_k)_{k\in\mathbb{N}}$ and $(y_k)_{k\in\mathbb{N}}$ are elements of S then for $\alpha, \beta \in \mathbb{R}$ we get

$$\lim_{n\to\infty} \frac{\sum_{k=1}^{n} \alpha x_k + \beta y_k}{n} = \lim_{n\to\infty} \frac{\sum_{k=1}^{n} \alpha x_k}{n} + \lim_{n\to\infty} \frac{\sum_{k=1}^{n} \beta y_k}{n} =$$

$$\alpha \lim_{n\to\infty} \frac{\sum_{k=1}^{n} x_k}{n} + \beta \lim_{n\to\infty} \frac{\sum_{k=1}^{n} y_k}{n}, \tag{6.4.11}$$

which means that S is a vector subspace of $\mathbb{R}^{\mathbb{N}}$.

We have to show that S is Borel subset of $\mathbb{R}^{\mathbb{N}}$.

For $i \in \mathbb{N}$, we denote by Pr_i the ith projection on $\mathbb{R}^{\mathbb{N}}$ defined by

$$Pr_i((x_k)_{k\in\mathbb{N}}) = x_i \tag{6.4.12}$$

for $(x_k)_{k\in\mathbb{N}} \in \mathbb{R}^{\mathbb{N}}$.

We put $S_n = \frac{\sum_{i=1}^{n} Pr_i}{n}$ for $n \in \mathbb{N}$. Then the set of all infinite samples $x \in \mathbb{R}^{\mathbb{N}}$ for which there exists a finite limit $\lim_{n\to\infty} S_n(x)$ coincides with S. On the other hand, taking into account that $S_n : \mathbb{R}^{\mathbb{N}} \to \mathbb{R}$ is a continuous function for $n \in \mathbb{N}$ and the equality

$$S = \cap_{p=1}^{\infty} \cup_{n=1}^{\infty} \cap_{q=n}^{\infty} \cap_{m=1}^{\infty} \{x : x \in \mathbb{R}^{\mathbb{N}} \ \& \ |S_{q+m}(x) - S_q(x)| \leq 1/p\} \quad (6.4.13)$$

holds, we claim that S is Borel subset of $\mathbb{R}^{\mathbb{N}}$.

We put $v = (1, 2, 3, \ldots)$. Let us show that v spans a line L such that every translate of L meets S in at most one point; in particular, L is a probe for the complement of S. Indeed, assume the contrary. Then there will be an element $(z_k)_{k\in\mathbb{N}} \in \mathbb{R}^{\mathbb{N}}$ and two different parameters $t_1, t_2 \in \mathbb{R}$ such that $(z_k)_{k\in\mathbb{N}} + t_1(1, 2, 3, \ldots) \in S$ and $(z_k)_{k\in\mathbb{N}} + t_2(1, 2, 3, \ldots) \in S$. Because S is a vector space we deduce that $(t_2 - t_1)(1, 2, 3, \ldots) \in S$. Using the same argument we claim that $(1, 2, 3, \ldots) \in S$ because $t_2 - t_1 \neq 0$, but the latter relation is false because

$$\lim_{n\to\infty} \frac{\sum_{k=1}^{n} k}{n} = \lim_{n\to\infty} \frac{n+1}{2} = +\infty. \quad (6.4.14)$$

This ends the proof of the theorem.

By the scheme presented in the proof of Theorem 6.4.3, we can get the validity of the following assertion.

Theorem 6.4.4 *Let $T = \overline{\lim}\widetilde{T}_n$ be the statistic from Lemma 6.3.1. Then for "almost every" infinite sample the statistic T does not exist.*

Theorem 6.4.5 *Let $T : \mathbb{R}^{\mathbb{N}} \to \mathbb{R}$ be an infinite sample statistic defined by*

$$T((x_k)_{k\in\mathbb{N}}) = \lim_{n\to\infty} \frac{\sum_{k=1}^{n} x_k}{\sqrt{n}\sigma}, \quad (6.4.15)$$

where $\sigma > 0$. Then for "almost every" infinite sample the statistic T does not exist.

Proof Let S be defined by

$$S = \{(x_k)_{k\in\mathbb{N}} : (x_k)_{k\in\mathbb{N}} \in \mathbb{R}^{\mathbb{N}} \ \& \text{ there exists a finite limit } \lim_{n\to\infty} \frac{\sum_{k=1}^{n} x_k}{\sqrt{n}\sigma}\}. \quad (6.4.16)$$

It is obvious that S is a vector subspace of $\mathbb{R}^{\mathbb{N}}$. Indeed, if $(x_k)_{k\in\mathbb{N}}$ and $(y_k)_{k\in\mathbb{N}}$ are elements of S then for $\alpha, \beta \in \mathbb{R}$ we get

$$\lim_{n\to\infty} \frac{\sum_{k=1}^{n} \alpha x_k + \beta y_k}{\sqrt{n}\sigma} = \lim_{n\to\infty} \frac{\sum_{k=1}^{n} \alpha x_k}{\sqrt{n}\sigma} + \lim_{n\to\infty} \frac{\sum_{k=1}^{n} \beta y_k}{\sqrt{n}\sigma} =$$

$$\alpha \lim_{n\to\infty} \frac{\sum_{k=1}^{n} x_k}{\sqrt{n}\sigma} + \beta \lim_{n\to\infty} \frac{\sum_{k=1}^{n} y_k}{\sqrt{n}\sigma}, \quad (6.4.17)$$

which means that S is a vector subspace of $\mathbb{R}^{\mathbb{N}}$.

We have to show that S is Borel subset of $\mathbb{R}^{\mathbb{N}}$.

For $i \in \mathbb{N}$, we denote by Pr_i the ith projection on $\mathbb{R}^{\mathbb{N}}$ defined by

$$Pr_i((x_k)_{k\in\mathbb{N}}) = x_i \quad (6.4.18)$$

for $(x_k)_{k\in\mathbb{N}} \in \mathbb{R}^{\mathbb{N}}$.

We put $S_n = \frac{\sum_{i=1}^{n} Pr_i}{\sqrt{n}\sigma}$ for $n \in \mathbb{N}$. Then, on the one hand, the set of all infinite samples $x \in \mathbb{R}^{\mathbb{N}}$ for which there exists a finite limit $\lim_{n\to\infty} S_n(x)$ coincides with S. However, taking into account that $S_n : \mathbb{R}^{\mathbb{N}} \to \mathbb{R}$ is a continuous function for $n \in \mathbb{N}$ and the equality

$$S = \cap_{p=1}^{\infty} \cup_{n=1}^{\infty} \cap_{q=n}^{\infty} \cap_{m=1}^{\infty} \{x : x \in \mathbb{R}^{\mathbb{N}} \ \&\ |S_{q+m}(x) - S_q(x)| \le 1/p\} \quad (6.4.19)$$

holds, we claim that S is a Borel subset of $\mathbb{R}^{\mathbb{N}}$.

We put $v = (1, 2, 3, \ldots)$. Let us show that v spans a line L such that every translate of L meets S in at most one point; in particular, L is a probe for the complement of S. Indeed, assume the contrary. Then there will be an element $(z_k)_{k\in\mathbb{N}} \in \mathbb{R}^{\mathbb{N}}$ and two different parameters $t_1, t_2 \in \mathbb{R}$ such that $(z_k)_{k\in\mathbb{N}} + t_1(1, 2, 3, \ldots) \in S$ and $(z_k)_{k\in\mathbb{N}} + t_2(1, 2, 3, \ldots) \in S$. Because S is a vector space we deduce that $(t_2 - t_1)(1, 2, 3, \ldots) \in S$. Using the same argument we claim that $(1, 2, 3, \ldots) \in S$ because $t_2 - t_1 \neq 0$, but the latter relation is false because

$$\lim_{n\to\infty} \frac{\sum_{k=1}^{n} k}{\sqrt{n}\sigma} = \lim_{n\to\infty} \frac{(n+1)\sqrt{n}}{2\sigma} = +\infty. \quad (6.4.20)$$

This ends the proof of the theorem.

Remark 6.4.2 Note that Theorems 6.4.3 and 6.4.5 contain interesting information (in terms of "prevalence") about the structures of domains of some infinite sample statistics that help us well explain the meaning of Jacob Cohen's conjecture [Coh] which asserted that *null hypothesis significance testing does not exist.*

Remark 6.4.3 Note that main results of Chap. 6 were obtained in [PK].

References

[Ch] Chow, S.L.: Statistical Significance: Rationale, Validity and Utility (1997)

[Ch1] Christensen, J.R.: Measure theoretic zero sets in infinite dimensional spaces and applications to differentiability of Lipschitz mappings. Actes du Deuxieme Colloque d'Analyse Fonctionnelle de Bordeaux (Univ. Bordeaux). **I**, 29–39 (1973). Publ. Dep. Math. (Lyon) **10**(2), 29–39 (1973)

[Coh] Cohen, J.: The earth is round ($p < .05$). Am. Psychol. **49** (12), 9971003 (1994)

[D] Dougherty, R.: Examples of non-shy sets. Fund. Math. **144**, 73–88 (1994)

[HSY] Hunt, B.R., Sauer, T., Yorke, J.A.: Prevalence: a translation-invariant "Almost Every" on infinite-dimensional spaces. Bull. (New Ser.) Am. Math. Soc. **27**(2, 10), 217–238 (1992)

[IS] Ibramkhallilov, I.S., Skorokhod, A.V.: On well-off estimates of parameters of stochastic processes (in Russian), Kiev (1980)
[Kh] Kharazishvili, A.B.: Topological Aspects of Measure Theory (in Russian). Naukova Dumka, Kiev (1984)
[K] Kline, R.: Beyond significance testing: reforming data analysis methods in behavioral research. American Psychological Association, Washington, DC (2004)
[LMS] Harlow, L.L., Mulaik, S.A., Steiger, J.H.: What if there were no significance tests? Lawrence Erlbaum Associates (1997)
[MZ] McCloskey, D.N., Ziliak, S.T.: The Cult of Statistical Significance: How the Standard Error Costs Us Jobs, Justice, and Lives, University of Michigan Press (2008)
[MH] Morrison, D., Henkel, R.: The Significance Test Controversy, AldineTransaction (2006)
[M] Mycielski, J.: Some unsolved problems on the prevalence of ergodicity, instability, and algebraic independence. Ulam Q. **1**(3) 30 ff. approx. 8 pp. (1992)
[N] Nunnally, J.: The place of statistics in psychology. Educ. Psychol. Meas. **20**(4), 641–650 (1960)
[O] Oakes, M.: Statistical Inference: A Commentary for the Social and Behavioural Sciences. Wiley, Chichester (1986)
[PK] Pantsulaia, G., Kintsurashvili, M.: Why is null hypothesis rejected for "almost every" infinite sample by some hypothesis testing of maximal reliability? J. Stat.: Adv. Theory Appl. **11**(1), 45–70 (2014)
[ZPS] Zerakidze, Z., Pantsulaia, G., Saatashvili, G.: On the separation problem for a family of Borel and Baire G-powers of shift-measures on R. Ukr. Math. J. **65**(4), 470–485 (2013)

Appendix A
Curriculum Vitae

CV of Professor Gogi Rauli Pantsulaia

Name: Gogi Rauli Pantsulaia
Born: September 14, 1960, Georgia
Citizenship: Georgia
Address: Department of Mathematics, Georgian Technical University, 77 Kostava st. Tbilisi 0143, Georgia
Tel: Office 00 995 32 36 47 90
Home 00 995 32 60 08 68
E-mail address: g.pantsulaia@gtu.ge

Education

1974 - Finished 8 class at I. Gogebashvili Sukhumi School No.11.
1977 - Finished V. Komarov Tbilisi Secondary School No. 2, with physics and mathematics orientation
1982 - MSc in Mathematics. Tbilisi State University, Georgia. Supervizor - Doctor of Physics and Mathematics, Professor G. Nijaradze, Thesis title: "Vitali systems and differentiability of absolutely continuous functions". Awarded Gold Diploma (highest grade for thesis and course work)
1985 - Certificate of completion of three years PhD course in Theory of probability and Mathematics statistics. Tbilisi State University. Supervisors: Doctor of Physics and Mathematics, Professor Gvandji Mania. Doctor of Physics and Mathematics, Professor Kharazishvili Alexander;
1985 - PhD in Mathematics. Institute of mathematics, Ukrainian Academy of Sciences, Kiev, Ukraine; Leading organization: Steklov Institute of Mathematics, Moscow. Thesis title: "Some properties of probability measures in space

G. Pantsulaia, *Applications of Measure Theory to Statistics*,
DOI 10.1007/978-3-319-45578-5

with additional structures" (01.01.05-Probability Theory). Supervisor: Doctor of Physics and Mathematics, Professor Kharazishvili Alexander

2003 - Degree of Doctor of Physics and Mathematics, Iv. Javakhishvili Tbilisi State University, I. Vekua Institute of Applied Mathematics, Tbilisi, Georgia (University st.2, Tbilisi, 0143) Thesis title: "On Some Quasiinvariant Measures and Dynamical Systems in Infinite-Dimensional Vector Spaces" (01.01.08-Mathematical Cybernetics), Scientific consultant-Doctor of Physics and Mathematics, Professor Kharazishvili Alexander.

Work Experience

1985–1987 - Research Fellow, Institute of the Management System, Academy of Georgian Sciences.

1988–2004 - Associate professor, Department of Mathematics, Georgian Technical University.

1997–2004 - Invited professor, Sukhumi branch of I. Javakhishvili Tbilisi State University.

2004–2006 - Professor at Department of Mathematics of Georgian Technical University.

2005–2006 - Senior researcher at Division of Discrete Mathematics of I. Vekua Institute of Applied Mathematics of Tbilisi State University.

2006–to day - Full professor at Division of Mathematical Analysis of Department of Mathematics of Georgian Technical University.

2009–to day - Senior researcher at Division of Discrete Mathematics of I. Vekua Institute of Applied Mathematics of Tbilisi State University.

Scientific Interest and Achievements

Set-theoretical aspects of real analysis, measure theory and related areas of mathematics.

Text Books and Booklets

1. G. Pantsulaia, Probability theory and mathematical statistics, part I, Tbilisi State University Publishers, © Copyright 1997-1998 (ISBN 99928.56-65-3) (1998) (in Georgian) (textbook)
2. G. Pantsulaia, B. Misabishvili, Invariant Borel measures in the vector space $R^{[0,1]}$, Publishers "Intelekti", © Copyright 2000 (ISBN 99928-860-3 x), (2000) (booklet)
3. G. Pantsulaia, B. Misabishvili,On some properties of nonmeasurable functions defined on metric spaces, Publishers "Intelekti", © Copyright 2000 (ISBN 99928-860-5 –6), (2000), (booklet)
4. G. Pantsulaia, L. Khocholava, Homogeneous dynamical systems and their properties, Publishers "Intelekti", © Copyright 2005 (ISBN 99928-860-4 –8), (2000) (booklet)

5. G. Pantsulaia, Elements of probability theory, Georgian Technical University Publishers, © Copyright 1004-2005 (ISBN 99940-40-33-2), (2005), (in Georgian)(textbook)
6. G. Pantsulaia, Elements of probability theory, Georgian Technical University Publishers, © Copyright 2005 (ISBN 99940-4809-0), (2005), (in Georgian)(textbook)

Monographs

1. G. Pantsulaia, ***Invariant and Quasiinvariant Measures in Infinite-Dimensional Topological Vector Spaces,*** Nova Science Publishers, © Copyright 2004-2007 (ISBN.1-60021-719-2), New York -2007, xii+231 pp.
   Web page:
   https://www.novapublishers.com/catalog/product_info.php?products_id=5715
2. G. Pantsulaia, ***Generators of Shy Sets in Polish Groups***. Nova Science Publishers, Inc., New York, 2011, xii -227 p (**ISBN:** 978-1-61728-030-6)
   Web page:
   https://www.novapublishers.com/catalog/product_info.php?products_id=14884
3. G. Pantsulaia, ***Selected topics of an infinite-dimensional classical analysis***. Nova Science Publishers, Inc., New York, 2012, xii -183 p (**ISBN:** 1619427176, **ISBN**: 9781619427174).
   Web page:
   https://www.novapublishers.com/catalog/product_info.php?products_id=31475
4. G. Pantsulaia, ***Selected topics of invariant measures in Polish groups***. Nova Science Publishers, Inc., New York *(2013)*, xii - 210 p. (**ISBN:** 978-1-62948-831-8)
   https://www.novapublishers.com/catalog/product_info.php?products_id=47326

Participation in International Conferences

1. ***Enlarged Sessions of the Seminar of I. Vekua Inst. Appl. Math***, Tbilisi (Georgia), 1994.
   Web page: http://www.viam.hepi.edu.ge/enlses/vo19.htm
2. ***ISPM, School and Colloguium: Stochastic Analysis and Applications in Control, Statistics and Financial Modeling***, Tbilisi (Georgia), September 1–7, 2002.
   Web page: http://www.rmi.acnet.ge/ispm/ispm-02/stat/abs.htm
3. ***Conference "Kolmogorov-100". MSU. Sect. 1. Dynamical Systems and Ergodic Theory.*** Moskow (Russia), 15–21 June, 2003.
   Web page: http://kolmogorov100.mi.ras.ru/info91.htm
4. ***Conference in Probability Theory and Mathematical statistics Dedicated to Centary of A.N. Kolmogorov,*** Tbilisi (Georgia), September 21–27, 2003.
   Web page: http://www.rmi.acnet.ge/ispm/ispm-02/stat/welcome.htm
5. ***International Conference: Modern Problems and New Trends in Probability Theory,*** Chernivtsi (Ukraine), June 19–26, 2005.
6. ***International School and Workshop on Function Spaces***, Integral Transforms and Applications in PDE, Dedicated to the 100th Birthday Anniversary of Professor Archil Kharadze, Tbilisi (Georgia), August 31–September 5, 2005.

Web page: http://www.rmi.acnet.ge/ispm/ispm-05/func/program.html#programme

7. ***International Conference***, Skorokhod Space, 50 Years on, Kyiv (Ukraine), 17–23 June, 2007.
8. ***International Workshop on Variable Exponent Analysis and Related Topics***, Tbilisi (Georgia) September 2–5, 2008.
   Web page: http://www.rmi.acnet.ge/eng/VEART/WEBSITE.pdf
9. ***International conference on Modern Problems in Applied Mathematics*** Dedicated to the 90th Anniversary of the Iv. Javakhisvili Tbilisi State University and 40th Anniversary of the I. Vekua Institute of Applied Mathematics, Section: Foundation of Mathematics and Mathematical Logic, Tbilisi (Georgia), 7–9 October, 2008.
   Web page: http://www.viam.science.tsu.ge/viam40/tezisebis_krebuli.pd
10. ***37th Winter School in Abstract Analysis***, Kacov (Czech), 17–22 January, 2009.
    Web page: http://www.karlin.mff.cuni.cz/~lhota/
11. **Stochastic analysis and random dynamics, International conference,** Lviv (Ukraine), 14. VI–20.VI, 2009
    Web page: http://www.imath.kiev.ua/~sard/program2009.rtf
12. ***39th Winter School in Abstract Analysis***, Kacov (Czech), 15–24 January, 2011.
    Web page: http://www.karlin.mff.cuni.cz/~lhota/
13. ***Second Annual Conference of the Georgian Mathematical Union,*** September 15–19, 2011, Batumi
    Web page: http://www.rmi.ge/~gmu/PDF_files/Engl-FIRST-Ann-Bat-2011.pdf
    Gogi Pantsulaia, On Witsenhausen–Kalai Constants for a Gaussian Measure on the Infinite-Dimensional Complex-Valued Unite Sphere, II International conference of the Georgian Mathematical Union, Book of Abstracts, Georgian Mathematical Union, Batumi, Georgia, September 15–19, 2011, Batumi (2012), p. 60
14. 40***th Winter School in Abstract Analysis***, Klenci (Czech), 14–21 January, 2012.
    Web page: http://www.karlin.mff.cuni.cz/~lhota/
    Gogi Pantsulaia and Givi Giorgadze, On some applications of infinite dimensional cellular matrices, 40th Winter School In Abstract Analysis, 14–21 January, 2012 (Report)
15. ***Third annual conference of the Georgian Mathematical Union,*** *September 2–9, 2012, in. Batumi, Black Sea Resort of Georgia.*
    Gogi R. Pantsulaia, Givi P. Giorgadze, A description of behaviors of some phase motions in R^{∞} in terms of ordinary and standard "Lebesgue measures", III International conference of the Georgian Mathematical Union, Book of Abstracts, Georgian Mathematical Union, Batumi, Georgia, September 2–9 (2012), p. 75
16. **International conference dedicated to 120-th anniversary of Stefan Banach** September 17–21, 2012 **Ivan Franko National University of Lviv,** Lviv, Ukraine
    Web page: http://atlas-conferences.com/cgi-bin/abstract/cbeg-96

Gogi R. Pantsulaia, Zurab Zerakidze, Gimzer Saatashvili, On consistent estimators of a useful signal in the linear one-dimensional stochastic model when an expectation of the transformed signal is not defined, International conference dedicated to 120-th anniversary of Stefan Banach, Abstracts of Reports, Lviv, Ukraine, September 17–21 (2012), p. 274.

17. The First Scientific Conference in Exact and Natural Sciences, Ivane Javakhishvili Tbilisi State University, Institute of Statistical Research, Tbilisi (Georgia), January 22–26, 2013
Gogi Pantsulaia, Zurab Zerakidze and Gimzer Saatashvili, On consistent estimators of a useful signal in the linear one-dimensional stochastic model when an expectation of the transformed signal does not exist
18. Second International Conference on "Modern Problems in Applied Mathematics" Dedicated to the 95th Anniversary of I. Javakhishvili Tbilisi State University and 45th Anniversary of I.Vekua Institute of Applied Mathematics
Gogi Pantsulaia, On a certain problem of H. Shi in Solovay's model, Conferences Organized by VIAM TSU, 4–7 September (2013), Tbilisi
19. Conferences Organized by The Georgian Mathematical Union, September 9-15 (2013), IV Annual Conference of the Georgian Mathematical Union Dedicated to Victor Kupradze on his 110-th birthday anniversary.
Gogi Pantsulaia and Nino Rusiashvili, On a certain version of Erdos problem
20. ***The 42th Winter School in Abstract Analysis***, January 11–18, 2014 (Hotel Mánes) Svratka.
Gogi Pantsulaia, On some problems of abstract analysis
Web page: http://www.karlin.mff.cuni.cz/~lhota/.

Membership of Editorial Boards of International Scientific Journals

1. *Georgian International Journal of Science and Technology,*
2. *Intelectual.*

Reviewer of International Scientific Journals

1. Mathematical Reviews
2. *Georgian Mathematical Journal,*
3. *Mathematical Analysis and Applications,*
4. *Proc. A. Razmadze Math. Inst.*
5. *Georg. Inter. J. Sci. Tech., Nova Science Publishers,*
6. *"Intelectual".*

Membership of Mathematical Societies

1. Member of Georgian Mathematical Society
2. Member of American Mathematical Society.

Participation in Scientific Grant Projects

2005	Georgian Ministery of Education and Science	**Grand /GNSF/ST 01/01-90.**
2008–		
2009	Georgian National Foundation	**Grand CNSF/ST 07/3-178**
2009	Georgian National Foundation	**Grand CNSF/ST 08/3-391**
2010	Shota Rustaveli National Foundation	**Grand CNSF/ST 09/3-105**
2013		
2013	Shota Rustaveli National Foundation	**Grand CNSF/ST 31-24**
2014		
2013	Shota Rustaveli National Foundation	**Grand CNSF/ST 31-25**
2016		
2015	Shota Rustaveli National Foundation	**Grand FR/116/5-100/14**
2018		

Scientific Publications of Gogi Rauli Pantsulaia

[1] G. Pantsulaia, On Generalized Integrals for Vector-Functions, Theses of Scientific *Reports of the Republic School-Seminar "Topological Aspects of Functions Theory"*, Tbilisi:1984 (in Russian).

[2] G. Pantsulaia, Some properties of families of probability measures, *Soobshch. Akad. Nauk Gruzin. SSR* 120 (1985), no. 2, 245–248 (1986) (in Russian).

[3] G. Pantsulaia, On the existence of a quasi-invariant measure on a nonlocally compact noncommutative topological group, *Soobshch. Akad. Nauk Gruzin. SSR* 120 (1985), no. 1, 53–55 (in Russian)

[4] G. Pantsulaia, Generalized integrals, *Soobshch. Akad. Nauk Gruzin. SSR* 117 (1985), no. 1, 33–36 (in Russian).

[5] G. Pantsulaia, A. Kirtadze, The property of essential uniqueness of invariant measures in vector spaces. (Russian) *Soobshch. Akad. Nauk Gruzin. SSR* 127 (1987), no. 1, 33–35

[6] G. Pantsulaia, Invariant and quasi-invariant measures in some function spaces. *Soobshch. Akad. Nauk Gruzin. SSR* 134 (1989), no. 3, part I, 489–492 (in Russian)

[7] G. Pantsulaia, Independent families of sets and some of their applications to measure theory. *Soobshch. Akad. Nauk Gruzin. SSR* 134 (1989), no. 1, 29–32 (in Russian).

[8] G. Pantsulaia, On orthogonal families of probability measures, *Trans. GPI,* (1989) 8(350), 106–112 (in Russian).

[9] G. Pantsulaia, On invariant measures in some functional spaces, *Trans. GTU,* (1991) no 10 (383), 9–14 (in Russian).

[10] A. Kirtadze, G. Pantsulaia, Invariant measures in the space $\mathbf{R}$ sp N. (Russian) *Soobshch. Akad. Nauk Gruzii* 141 (1991), no. 2, 273–276.

[11] G. Pantsulaia, A. Kirtadze, On uniquely definable extensions of invariant measures in the space $\mathbf{R}$ sp N. (Russian) *Soobshch. Akad. Nauk Gruzii* 149 (1994), no. 2, 189–192 (1995)

[12] G. Pantsulaia, On non-elementary extensions of the Haar measure, *Reports of Enlarged Sessions of the Seminar of I.Vekua Inst. Appl. Math.*, vol. 9. no 13, (1994), Tbilisi, p. 40–43.

[13] G. Pantsulaia, Density points and invariant extensions of the Lebesgue measure. (Russian) *Soobshch. Akad. Nauk Gruzii* 151 (1995), no. 2, 216–219 (1996). CMP 1 652 055.

[14] Pantsulaia G.R. The construction of invariant measures in the non-separable Banach space l∞ //Georgian Technical University Press. – 2000. – T. 430. – C. 18–20.

[15] G. Pantsulaia, B. Misabishvili, Invariant Borel measures in the vector space $R^{[0;\pi]}$, *Publishers "Intelekti", © Copyright 2000* (ISBN 99928-860-3 x), (2000) (booklet).

[16] G. Pantsulaia, B. Misabishvili, On some properties of nonmeasurable functions defined on metric spaces, *Publishers "Intelekti", © Copyright 2000* (ISBN 99928-860-5–6), (2000), (booklet).

[17] G. Pantsulaia, L. Khocholava, Homogeneous dynamical systems and their properties, *Publishers "Intelekti", © Copyright 2005* (ISBN 99928-860-4–8), (2000) (booklet).

[18] G. Pantsulaia, Vector fields of velocities in infinite-dimensional topological vector space R^N which preserve a measure μ, *Trans. of GTU,* 1(434) (2001), 9–16.

[19] G. Pantsulaia, On Economical Connection Property In Infinite-Dimensional Topological Vector Spaces, *Thesises of Reports, Georgian Mathematicians III Congress, Tbilisi, 11–13 Octomber* (2001), 100.

[20] G. Pantsulaia, Duality of measure and category in infinite-dimensional separable Hilbert space, *IJMMS* 30:6(2002). 353–363 PII. SO161171202012371.

[21] G. Pantsulaia, On the Strict Transitivity Property for Infinite-Dimensional Topological Vector Spaces, *Bull. Georgian Acad. Sci.*, 166, no 1 (2002), 32–35.

[22] G. Pantsulaia, On the Sierpinski-Erdos duality principle and Steinhaus property in infinite-dimensional topological vector spaces, *Bull. Georgian Acad. Sci.*, 166, no 2 (2002), 231–233.

[23] G. Pantsulaia, Representations of Borel Product-Measures by Haar Measure and Some of Their Applications, *Abstracts, ISPM, School and Colloguium: Stochastic Analysis and Applications in Control, Statistics and Financial Modeling, September 1–7, (2002), Tbilisi, Georgia.*

[24] G. Pantsulaia, Calculation of Jacobian for measurable linear operators in the Vector Space of all Real-Valued Sequences. *Theses of the Reports of Professors and Lecturers Scientific-Technical Conference Dedicated to the 80 Anniversary of Georgian technical University,* Tbilisi, (2002), 211.

[25] G. Pantsulaia, On some quasiinvariant measures and dynamical systems in infinite-dimensional vector spaces: Dis. for the doctor of phys. a. math. sciences 01.01.08 / Sci. consultant Alexander Kharazishvili; *Georg. techn. Univ., Dep. of math. no. 63 - Tb., 2003 - [3], 182p. – Bibliogr.: p. 172–182*

[26] G. Pantsulaia, On separation properties for families of probability measures, *Georgian Mathematical Journal*, vol. 10, no 2, (2003), 335–342.

[27] G. Pantsulaia, An Analogue of Lioville's Theorem in Infinite-Dimensional Separable Hilbert space l_2. *Collection of Abstracts (Comtributed Papers in English). Section 1. Dynamical Systems and Ergodic Theory. Conference "Kolmogorov-100". MSU.Moskow*. 15–21 June (2003), 79–80.

[28] G. Pantsulaia, Relations between Shy sets and Sets of ν_p-Measure Zero in Solovay's Model, *Bull. Polish Acad. Sci.*, vol. 52, No.1, (2004), 63–69.

[29] G. Pantsulaia, On an Invariant Borel Measure in Hilbert Space, *Bull. Polish Acad. Sci.*, vol. 52, No.2, (2004), 47–51.

[30] G. Pantsulaia, An Applications of independent families of sets to the measure extension problem, *Georgian Mathematical Journal*, vol. 11, no 2, (2004), 379–390.

[31] G. Pantsulaia, A. Kirtadze, On some notions of null sets in the infinite-dimensional topological vector space R^N. *Bull. of the Georgian Acad. Sci.*, 172 N2 (2005), 198–201.

[32] G. Pantsulaia, A. Kirtadze, Relation between shy sets and Haar null-Sets in the Banach Space l^∞, Bull. of the Georgian Acad. Sci., 172 N (3), (2005), 381–383.

[33] G. Pantsulaia, A. Kirtadze, On null sets in infinite dimensional separable Banach spaces International Conference: Modern Problems and New Trends in Probability Theory, *Abstracts II, Chernivtsi, Ukraine, June* 19–26), (2005) 56–57.

[34] G. Pantsulaia, On an infinite-dimensional analogy of Cramer Rule in Solovay model, *Thesises of Reports, Georgian Mathematicians IY Congress, Tbilisi*, 14–16, *November* (2005), 122

[35] G. Pantsulaia, A. Kirtadze, On null sets in infinite-dimensional Banach spaces. *Proc. A. Razmadze Math. Inst.* 139 (2005), 128–131.

[36] G. Pantsulaia, A. Kirtadze, Lebesgue nonmeasurable sets and the uniqueness of invariant measures in infinite-dimensional vector spaces. *Proc. A. Razmadze Math. Inst.* 143 (2007), 95–101.

[37] G. Pantsulaia, On an inductive limit of invariant measures, *Abstracts, International Conference, Skorokhod Space, 50 Years on, 17–23 June, Kyiv, Ukraine,* (2007) 86–88.

[38] G. Pantsulaia, Invariant and quasiinvariant measures in infinite-dimensional topological vector spaces. *Nova Science Publishers, Inc., New York*, 2007. xii+234 pp.

[39] G. Pantsulaia, On intersection of problems of theoretical statistics and set theory, *Far East Journal of Theoretical Statistics*, Volume 25, Issue 2, Pages 251–271 (July 2008)

[40] G. Pantsulaia, On Komjat's conjecture in the Martin-Solovay model, *Georg. Inter. J. Sci. Tech., Nova Science Publishers*, 1 (1) (2008), 71–76

[41] G. Pantsulaia, On a Riemann Integrability of functions defined on infinite-dimensional rectangles *Sokhumi state university proceedings, Mathematics and computer series, v. 5 (2008),*

[42] G. Pantsulaia, G.Giorgadze, On some applications of generators of shy-sets in infinite-dimensional analysis, International workshop on variable exponent analysis and related topics, September, 2–5 (2008) p. 1–46 (website: http://haota.ualg.pt/vea/Pantsulaia.pdf)

[43] G. Pantsulaia, A Relation Between Some Translation-Invariant Borel Measures On R^N, Book of abstracts, *International Conference on Modern Problems in Applied Mathematics Dedicated to the 90th Anniversary of the Iv. Javakhishvili Tbilisi State University and 40th Anniversary of the I. Vekua Institute of Applied Mathematics 7–9 October, 2008, Tbilisi, p. 14*

[44] G. Pantsulaia, G.Giorgadze, On analogy of Liouville theorem in Hilbert space l_2, Georg. Inter. J. Sci. Tech., Nova Science Publishers, Volume 1, Issue 2, (2008), 167–179. *Georgian Science and Technology Developments (Editor F. Columbus)* (2011), 293–306

[45] G. Pantsulaia, On Left-Invariant Probability Measures on General Groups, *Georg. Inter. J. Sci. Tech., Nova Science Publishers*, 1(3) (2008), 209–226. *Georgian Science and Technology Developments (Editor F. Columbus)* (2011), 307–324

[46] G. Pantsulaia, On generators of shy sets on Polish topological vector spaces, *New York J. Math.*,14 (2008), 235–261

[47] G. Pantsulaia, On a certain partition of the non-locally compact abelian Polish group $\mathbb{R}$ sp N. *Proc. A. Razmadze Math. Inst.* 149 (2009), 75–86

[48] G. Pantsulaia, Change of variable formula for "Lebesgue measures" on $\mathbb{R}$ sp N., *J. Math. Sci. Adv. Appl.* 3, 2 (1) (2009), 1–12

[49] G. Pantsulaia, On a generalized Fourier μ-series in some infinite-dimensional Polish topological vector spaces, *Georg. Inter. J. Sci. Tech., Nova Science Publishers*, 1(4) (2008), 313–326. *Georgian Science and Technology Developments (Editor F. Columbus)* (2011), 355–368

[50] G. Pantsulaia, On a certain example of a non-locally compact non-abelian Polish group which does not satisfy CCC, *Sokhumi state university proceedings, Mathematics and computer series, v. 4 (2008), 94–98*

[51] G. Pantsulaia, G. Giorgadze, On Lioville type theorems for Mankiewicz and Preiss-Tišer generators in R^N, *Georg. Inter. J. Sci. Tech., Nova Science Publishers*, Volume 2, Issue 1 (2009) 89–101. *Focus on Science and Technology from a Georgian Perspective* (Editor S. Gotsiridze) (2011), 89–102

[52] G. Pantsulaia, On infinite version of some classical results in Linear Algebra and Vector Analysis, *Georg. Inter. J. Sci. Tech., Nova Science Publishers*, Volume 2, Issue 1, (2009), 73–88. *Focus on Science and Technology from a Georgian Perspective* (Editor S.Gotsiridze) (2011), 73–87

[53] G. Pantsulaia, On a certain transition kernel in Solovay model, *Fifth congress of mathematicians of Georgia, Batumi-Kutaisi (Georgia), Abstracts of contributed talks*, p. 31(2009).

[54] G. Pantsulaia, On a certain property of zeroes of the Riemann's extended zeta function, *J. of Algebras, Groups and Geometries*, 26, No. 2, 223–229 (2009).

[55] G. Pantsulaia, On a certain problem of E. Szpilrajn for non-separable complete metric spaces, *Abstracts, stochastic analysis and random dynamics, International conference*, June 14–20, Lviv, Ukraine, 190–191 (2009).

[56] G. Pantsulaia, On ordinary and Standard Lebesgue Measures on R^{∞}, *Bull. Polish Acad. Sci.* 73(3) (2009), 209–222

[57] G. Pantsulaia, On Lioville type theorems for generators of shy sets in the infinite-dimensional Polish topological vector spaçe R^N, *Abstracts, stochastic analysis and random dynamics, International conference*, June 14–20, Lviv, Ukraine, 192–193 (2009).

[58] G. Pantsulaia, On a certain criterion of shyness for subsets in the product of unimodular Polish groups that are not compact, *J. Math. Sci. Adv. Appl.*, 3 (2) (2009), 287–302

[59] G. Pantsulaia, On Erdosh one problem for extensions of Lebesgue measures, *Hadronic J.*, Volume 32, Number 4, August (2009), 425–433

[60] G. Pantsulaia, On T-shy sets in Radon metric groups, *J. Math. Sci. Adv. Appl.* 5 (1) (2010), 149–186

[61] G. Pantsulaia, An expansion into infinite-dimensional multiple trigonometric series of a square integrable function in R^{∞}, *Georg. Inter. J. Sci. Tech., Nova Science Publishers*, Volume 2, Issue 4 (2010)(# GNSF = ST 08/3-391, # GNSF = ST 09_144-3-105)

[62] G. Pantsulaia, On Gaussian measures on $\mathbf{R}^I$ and some of their applications, *Georg. Inter. J. Sci. Tech., Nova Science Publishers*, Volume 2, Issue 4 (2010).

[63] G. Pantsulaia, On phase flows in $\mathbf{R}^{\infty}$ defined by Riemann's alternating zeta function, *Algebras Groups Geom.* 27 (2010) **179–190** (# GNSF / ST 09/3 – 105)

[64] Pantsulaia, G.; Zhivkov, N.V. On Witsenhausen-Kalai constants for extensions of the surface measure on S sp n. *Algebras Groups Geom.* **27**, (2010), no. 4, 391–407.

[65] G. Pantsulaia, On Planar Sets Without of Vertices of a Triangle of an Area One, *Georg. Inter. J. Sci. Tech., Nova Science Publishers*, Volume 2, Issue 2 (2010), 103–107. *Focus on Science and Technology from a Georgian Perspective* (Editor S. Gotsiridze) (2011), 103–108

[66] G. Pantsulaia, On a Certain Partition of the Lebesgue Null Set in R^n, *Georg. Inter. J. Sci. Tech., Nova Science Publishers*, Volume 2, Issue 2 (2010), 109–117. *Focus on Science and Technology from a Georgian Perspective* (Editor S. Gotsiridze) (2011), 109–117

[67] G. Pantsulaia, On a Certain Application of Preiss-Tiser Generators, *Georg. Inter. J. Sci. Tech., Nova Science Publishers*, Volume 2, Issue 2 (2010), 119–123. *Focus on Science and Technology from a Georgian Perspective* (Editor S. Gotsiridze) (2011), 119–124

[68] G. Pantsulaia, On Measurability of Unions of Plane Disks, *Georg. Inter. J. Sci. Tech., Nova Science Publishers*, Volume 2, Issue 2 (2010), 125–129. *Focus on Science and Technology from a Georgian Perspective* (Editor S. Gotsiridze) (2011), 125–129

[69] G. Pantsulaia, On a standard product of an arbitrary family of σ-finite Borel measures with domain in Polish spaces, *Theory Stoch. Process, vol. 16(32), 2010, no. 1, p. 84–93*

[70] G. Pantsulaia, On a certain problem of Mauldin-Preiss-Weizsacker, *Rep. Enlarged Sess. Semin. I. Vekua Appl. Math.*, 24(2010), pp. 97–98

[71] G. Pantsulaia, On ordinary and standard products of infinite family of σ -finite measures and some of their applications. *Acta Math. Sin.* (Engl. Ser.) 27 (2011), no. 3, 477–496

[72] G. Pantsulaia, On Uniform Measures in Banach Spaces, *Georg. Inter. J. Sci. Tech.*, Nova Science Publishers, Volume 2, Issue 3 (2010). *Technology from a Georgian Perspective* (Editor S. Gotsiridze) (2011), 239–247

[73] G. Pantsulaia, Riemann Integrability and Uniform Distribution in Infinite-Dimensional Rectangles, *Georg. Inter. J. Sci. Tech., Nova Science Publishers*, Volume 2, Issue 3 (2010). *Focus on Science and Technology from a Georgian Perspective* (Editor S. Gotsiridze) (2011), 249–260

[74] G. Pantsulaia, On Strict Standard and Strict Ordinary Products of Measures and Some of Their Applications, *Georg. Inter. J. Sci. Tech., Nova Science Publishers*, Volume 2, Issue 3 (2010). *Focus on Science and Technology from a Georgian Perspective* (Editor S. Gotsiridze) (2011), 261–294

[75] G. Pantsulaia, *Generators of Shy Sets in Polish Groups*. Nova Science Publishers, Inc., New York, 2011, xii–227 p

[76] G. Pantsulaia, G. Giorgadze, "Lebesgue "-null sets in $\mathbf{R}^\infty$ are not preserved under Lipschitz isomorphisms, *Georg. Inter. J. Sci. Tech., Nova Science Publishers*, Volume 3, Issue 1 (2011), 39–49(# GNSF = ST 09_144-3-105)

[77] G. Pantsulaia, G. Giorgadze, On invariant generators of shy sets for a certain nonhomogeneous stochastic differential equation in R^∞, *Georg. Inter. J. Sci. Tech., Nova Science Publishers*, Volume 3, Issue 1 (2011), 51–60.

[78] G. Pantsulaia, G. Giorgadze, On some applications of infinite-dimensional cellular matrices, *Georg. Inter. J. Sci. Tech., Nova Science Publishers*, Volume 3, Issue 1 (2011), 107–129.

[79] G. Pantsulaia, On singularity of non-σ-finite measures and transition kernels, *Georg. Inter. J. Sci. Tech., Nova Science Publishers*, Volume 3, Issue 1 (2011), 21–37.

[80] G. Pantsulaia, On uniformly distributed sequences of an increasing family of finite sets in infinite-dimensional rectangles, Real Anal. Exchange, Vol. 36(2) (2010/2011) 325–340

[81] G. Pantsulaia, On uniformly distribution in infinite-dimensional rectangles, *Georg. Inter. J. Sci. Tech., Nova Science Publishers*, Volume 3, Issue 2 (2011).

[82] G. Pantsulaia, G. Saatashvili, On separation of the family of N-powers of shift-measures in R^∞, *Georg. Inter. J. Sci. Tech.*, Volume 3, Issue 2 (2011), 189–195.

[83] G. Pantsulaia, On Invariant Extensions of the Standard Product of an Arbitrary Family of Haar Measures with Domain in Polish Groups, *Georg. Inter. J. Sci. Tech., Nova Science Publishers*, Volume 3, Issue 2 (2011).

[84] G. Pantsulaia, On a Certain Geometric Inequality on a Sphere S^n, *Georg. Inter. J. Sci. Tech., Nova Science Publishers*, Volume 3, Issue 2 (2011).

[85] G. Pantsulaia, On Convolution Problem on the Group $\{-1,+1\}^\infty$, *Georg. Inter. J. Sci. Tech., Nova Science Publishers*, Volume 3, Issue 3 (2011), 327–333

[86] G. Pantsulaia, On left-Invariant Borel Measures on the Product of Locally Compact Hausdorff Groups That Are Not Compact, *Georg. Inter. J. Sci. Tech., Nova Science Publishers*, Volume 3, Issue 3 (2011), 335–344

[87] G. Pantsulaia, *Selected topics of an infinite-dimensional classical analysis. Nova Science Publishers, Inc., New York*, 2012, xii–183.

[88] G. Pantsulaia, On an Equivalence of Outer Measures and Measures on Polish Spaces, *Georg. Inter. J. Sci. Tech., Nova Science Publishers*, Volume 3, Issue 4 (2012), 59–63.

[89] G. Pantsulaia, G. Giorgadze, On an explicit formula of a partial solution of non-homogeneous differential equation of the higher order with real constant coefficients, *Georg. Inter. J. Sci. Tech., Nova Science Publishers*, Volume 4, Issues 1,2 (2012), 37–47

[90] G. Pantsulaia, G. Giorgadze, Description of the behaviour of phase motions defined by generalized von Foerster-Lasota equations in R^∞ in terms of ordinary and standard "Lebesgue measures" (Dedicated to the memory of Professor Andrzej Lasota), *Georgian Int. J. Sci. Technol.* **4** (2012), no. 1–2, 47–62.

[91] Zurab Zerakidze, Gogi Pantsulaia, Gimzer Saatashvili, On consistent estimators of a useful signal in one dimensional stochastic model when the mean is not defined for transformed signal, *Georgian Int. J. Sci. Technol.* **4** (2012), no. 3-4, 189–198

[92] Tepper L. Gill, G.R. Pantsulaia, and W.W. Zachary, Constructive Analysis In Infinitely Many Variables, arXiv 1206-1764v2 [math-FA] 26 June 2012

[93] Tepper L. Gill, G.R. Pantsulaia, and W.W. Zachary, Constructive Analysis In Infinitely Many Variables, Communications in Mathematical Analysis, Volume 13, Number 1, pp. 107–141(2012) ISSN 19389787. http://arxiv.org/abs/1206.1764

[94] G. Pantsulaia, Tepper L. Gill, G. Giorgadze, On Dynamical Systems Defined by Partial Differential Equations of Infinite Order with Real Constant Coefficients, *Georg. Inter. J. Sci. Tech.*, Volume 4, Issues 3 (2013), pp. 53–94.

[95] G. Pantsulaia, On some examples of generators of shy sets in the Euclidean plane and related topics, *Georg. Inter. J. Sci. Tech.*, Volume 4, Issues 3 (2013), pp. 29–38

[96] G. Pantsulaia, On Uniformly Distributed Sequences on [-1/2,1/2], *Georg. Inter. J. Sci. Tech.*, Volume 4, Issues 3 (2013), pp. 21–27.

[97] G. Pantsulaia, On existence and uniqueness of generators of shy sets in Polish groups, *J. Math. Sci. Adv. Appl.* Volume no: 16, Issue no: 1–2, July and August (2012) pp. 61–78.

[98] T. Gill, A. Kirtadze, G. Pantsulaia, A. Plichko, The existence and uniqueness of translation invariant measure in separable Banach spaces, *Functiones et Approximatio*, 50.2 (2014), 1–19 doi:10.7169/facm/2014.50.2.

[99] Zurab Zerakidze, Gogi Pantsulaia, Gimzer Saatashvili, On the separation problem for a family of Borel and Baire G-powers of shift-measures on $\mathbb{R}$, *Ukrainian Mathematical Journal, v. 65, issue 4, 2013, 470-485.*
[100] Gogi Pantsulaia, On a Union fewer than c generalized shy sets, *Georgian Int. J. Sci. Technol.* 5 (2013), no. 1-2, 79–86
[101] Gogi Pantsulaia, Nino Rusiashvili, On questions of U. Darji and D. Fremlin, *Georg. Inter. J. Sci. Tech., Nova Science Publishers*, Volume 5, Number 3–4 (2013), 89–102.
[102] G. Pantsulaia, G. Giorgadze, On Generalized von Foersten-Lasota equations of the infinite order in R^{∞}, *Georg. Inter. J. Sci. Tech., Nova Science Publishers*, Volume 5, Number 3–4 (2013), pp 195–215
[103] G. Pantsulaia, A. Kirtadze, On a certain modification of P. Erdös problem for translation-invariant quasi-finite diffused Borel measures in Polish groups that are not locally compact, *Georg. Inter. J. Sci. Tech., Nova Science Publishers*, Volume 6, Issue 2 (2014),7–20 (# GNSF no 31/24, # GNSF no 31/25)
[103] T. Gill, A. Kirtadze, G. Pantsulaia, A. Plichko, N. Rusiashvili, On ordinary and standard "Lebesgue measures " in separable Banach spaces, *Georg. Inter. J. Sci. Tech., Nova Science Publishers*, Volume 5, Issues 3/4 (2013), 115–134
[104] G. Pantsulaia, On strange null sets in some vector spaces, Georg. Inter. J. Sci. Tech., Nova Science Publishers, Volume 6, Numbers 1, 2013(2014), pp. 41–48.
[105] Gogi Pantsulaia, Nino Rusiashvili, On a certain version of the Erdos problem, Georg. Inter. J. Sci. Tech., Nova Science Publishers, Volume 6, Number 3 (2014), pp. 67–74
[106] G. Pantsulaia, *Selected topics of invariant measures in Polish groups. Nova Science Publishers, Inc., New York (*2014), XI+210 p. ISSN 978-1-62948-831-8
[107] Gogi R. Pantsulaia, Tepper L.Gill, Givi P. Giorgadze, On a heat equation in an infinite-dimensional separable Banach space with Schauder basis, *Georg. Inter. J. Sci. Tech., Nova Science Publishers*, Volume 6, Number 2 (2014), p. 37–53
[108] G. Pantsulaia, M. Kintsurashvili, Why is Null Hypothesis rejected for "almost every" infinite sample by some Hypothesis Testing of maximal reliability?, *Journal of Statistics: Advances in Theory and Applications*, Volume 11, Number 1, 2014, Pages 45–70
[109] G. Pantsulaia and A. Kirtadze, On Witsenhausen-Kalai Constants for Infinite-Dimensional Surface Dynamical Measures, *Georg. Inter. J. Sci. Tech., Nova Science Publishers*, Volume 6, Number 2 (2014), p. 55–57
[110] Gogi R. Pantsulaia and Givi P. Giorgadze, On a Representation of the Solution of a Certain Generalized Heat Equation of Many Variables in a Multiple Trigonometric Series, *Georg. Inter. J. Sci. Tech., Nova Science Publishers*, Volume 6, Number 2 (2014), p. 21–36
[111] Gogi R. Pantsulaia, On Maximal Plane Sets Containing Only the Vertices of a Triangle with Area Less Than One, Georg. Inter. J. Sci. Tech., Nova Science Publishers, Volume 6, Number 2 (2014), p. 1–5.

[112] Gogi R. Pantsulaia, A classification of non-measurable real-valued functions defined on a metric space, *Georg. Inter. J. Sci. Tech., Nova Science Publishers*, Volume 6, Number 3 (2014), p. 59–66.

[113] G. Pantsulaia, M. Kintsurashvili, An effective construction of the strong objective infinite-sample well-founded estimate, *Proc. A. Razmadze Math. Inst.* 166 (2014), 113–119.

[114] G. Pantsulaia, M. Kintsurashvili, An objective infinite sample well-founded estimates of a useful signal in the linear one-dimensional stochastic model, *Rep. Enlarged Sess. Semin. I. Vekua Appl. Math.* 28 (2014), 90–93.

[115] Gogi Pantsulaia, Givi Giorgadze, A description of some phase motions in terms of ordinary and standard "Lebesgue measures" in R^{∞}, *Georg. Inter. J. Sci. Tech.*, Volume 6, Number 4 (2014), p. 285–329

[116] Murman Kintsurashvili, Tengiz Kiria, Gogi Pantsulaia, On objective and strong objective consistent estimates of unknown parameters for statistical structures in a Polish group admitting an invariant metric, Journal of Statistics: Advances in Theory and Applications, Volume 13, No. 2 (2015) 179–233

[117] Gogi Pantsulaia, Infinite-Dimensional Monte-Carlo Integration, *Monte Carlo Methods Appl.* **21** (2015), no. 4, 283–299.

[118] Gogi Pantsulaia, Solovay model and duality principle between the measure and Baire category in a Polish topological vector space H(X, S, μ), Reports of Enlarged Session of the Seminar of I. Vekua Institute of Applied Mathematics Volume 29, 2015, 103–106

[119] G. Pantsulaia, Equipment of sets with cardinality of the continuum by structures of Polish groups with Haar measures, International Journal of Advanced Research in Mathematics, *Sci. Press Ltd., Switzerland*, Vol. **5**, (2016) 8–22

[120] Gogi R. Pantsulaia and Givi P. Giorgadze, On a Linear Partial Differential Equation of the Higher Order in Two Variables with Initial Condition Whose Coefficients are Real-valued Simple Step Functions, J. Partial Diff. Eqs. Vol. **29**, (2016). No. 1, 1–13

[121] *M. Kintsurashvili, T. Kiria and G. Pantsulaia*, On Moore–Yamasaki–Kharazishvili type measures and the infinite powers of Borel diffused probability measures on R, Journal of Mathematical Sciences: Advances and Applications, Volume 38, 2016, Pages 73–82

[122] T. Kiria, G. Pantsulaia, Calculation of Lebesgue integrals by using uniformly distributed sequences, Transactions of A. Razmadze Mathematical Institute, Volume 170, Issue 3, December (2016), pp. 402–409

Index

G. Pantsulaia, *Applications of Measure Theory to Statistics*,
DOI 10.1007/978-3-319-45578-5

Printed in the United States
By Bookmasters